Thin Film Processes

Thin Film Processes

Special Issue Editor

Hyun Wook Jung

MDPI • Basel • Beijing • Wuhan • Barcelona • Belgrade

Special Issue Editor
Hyun Wook Jung
Department of Chemical and
Biological Engineering,
Korea University
Korea

Editorial Office
MDPI
St. Alban-Anlage 66
4052 Basel, Switzerland

This is a reprint of articles from the Special Issue published online in the open access journal *Processes* (ISSN 2227-9717) from 2018 to 2020 (available at: https://www.mdpi.com/journal/processes/special_issues/film_process).

For citation purposes, cite each article independently as indicated on the article page online and as indicated below:

LastName, A.A.; LastName, B.B.; LastName, C.C. Article Title. *Journal Name* **Year**, *Article Number*, Page Range.

ISBN 978-3-03936-264-6 (Pbk)
ISBN 978-3-03936-265-3 (PDF)

Contents

About the Special Issue Editor

Hyun Wook Jung received his PhD degree from Department of Chemical Engineering at Korea University in 1999. He spent his postdoc at Department of Chemical Engineering and Materials Science of University of Minnesota from 2000 to 2001. After working at the Applied Rheology Center (Engineering Research Center at Korea University) from 2001 to 2003, as a Research Professor, he joined Department of Chemical and Biological Engineering at Korea Univ. in 2003 as a faculty member. He has received the KU President Award 1993, the Bumseok Paper Award 2005 from the Korean Institute of Chemical Engineers, the Young Rheologist Award 2011 from the Korean Society of Rheology, the Best Paper Award 2015 from the Korean Federation of Science and Technology Societies, and several Granite-Tower Lecture Awards from Korea University. He was the former Editor-in-Chief of Korea-Australia Rheology Journal (2016–2019) and is currently the Associate Editor of Korean Journal of Chemical Engineering and the Editor of News and Information for Chemical Engineers. Dr. Jung's research fields include nonlinear dynamics/stability/sensitivity in rheological processes, fundamental coating and drying processes, microrheology using light scattering techniques, curing and crosslinking of organic coatings, micro flows, and particulate suspension flows.

Editorial

Special Issue on "Thin Film Processes"

Hyun Wook Jung

Department of Chemical and Biological Engineering, Korea University, Seoul 02841, Korea; hwjung@grtrkr.korea.ac.kr

Received: 8 May 2020; Accepted: 8 May 2020; Published: 11 May 2020

Thin film processes are significantly incorporated in manufacturing display panels, secondary batteries, fuel/solar cells, catalytic films, membranes, adhesives, and other commodity films. This Special Issue on "Thin Film Processes". listed recent progresses on thin film processes, covering theoretical considerations, experimental observations, and computational techniques.

Articles in this issue consider comprehensive studies on thin film processes and related materials. Moon et al. [1] proposed the novel fabrication method of liquid crystal devices (LCD). Single-wall carbon nanotubes (SWNTs) are deposited onto a substrate via layer-by-layer (LBL) process. This SWNT LBL electrode is an appealing replacement of conventional indium tin oxide (ITO) transparent electrode for its superior optoelectronic performance. Vertically aligned LCD cells with the new electrode exhibited a voltage-transmittance curve similar to that of the conventional ITO electrode, and the enhanced display performance with the number of SWNT layers. Tang et al. [2] investigated the film growth characteristics and anti-coking performance of CVD (chemical vapor deposition)-based TiN and TiO_2 anti-coking coatings. Coating growth characteristics are importantly dependent on deposition time, residence time, and partial pressure. Coking tests showed that both TiN and TiO_2 coatings satisfactorily inhibited catalytic coking. Li et al. [3] compared the performances of ultrafine powder coatings with two different fillers ($Al(OH)_3$ and $BaSO_4$). Powder coatings consist of polymer resins, fillers, curing agents, additives, and pigments. Unlike $BaSO_4$ filler, $Al(OH)_3$ filler contains the polar groups (hydrogen bond), which induces the chain formation between inorganic filler and organic polymer resin. Several tests were conducted to measure extensive mechanical properties of the films, and films with $Al(OH)_3$ fillers were found to be better in mechanical properties and durability aspects. Algahtani et al. [4] conducted a series of experiments to determine the most efficient deposition method for the ceramic surface coating with good anti-corrosion ability. SEM and optical microscope were used to observe microstructures of the coated films, and corrosion performance of the films was tested via several electrochemical experiments including inductive coupled plasma (ICP) test. Nguyen et al. [5] designed mathematical models for the drying of lemongrass using the convection drying device. Mass of lemongrass was measured under different drying conditions to calculate the moisture content, and moisture diffusivity and activation energy of the drying process were nicely fitted from raw data. Kern et al. [6] suggested an alternative approach for developing paint and coating products. The digital platform to stably suggest the new formulation of paints from reliable data was ingeniously established. In this experiment, the digital platform effectively reduced the non-value-added activities and also the total throughput time. Gathering reliable safety and/or price data and formulating them into the structured form are crucial in this method. Khan et al. [7,8] and Ullah et al. [9] focused on the thin film flows on time-dependent stretching surface under the presence of the magnetic field. This magnetohydrodynamics (MHD) issue was scrutinized both numerically and analytically, giving good agreement with each other. The analytical results were acquired employing the homotopy analysis method (HAM).

Conflicts of Interest: The author declares no conflict of interest.

References

1. Moon, G.; Jang, W.; Son, I.; Cho, H.; Park, Y.; Lee, J. Fabrication of New Liquid Crystal Device Using Layer-by-Layer Thin Film Process. *Processes* **2018**, *6*, 108. [CrossRef]
2. Tang, S.; Liu, T.; Duan, S.; Guo, J.; Tang, A. Comparison of Growth Characteristics and Properties of CVD TiN and TiO$_2$ Anti-Coking Coatings. *Processes* **2019**, *7*, 574. [CrossRef]
3. Li, W.; Franco, D.; Yang, M.; Zhu, X.; Zhang, H.; Shao, Y.; Zhang, H.; Zhu, J. Comparative Study of the Performances of Al(OH)$_3$ and BaSO$_4$ in Ultrafine Powder Coatings. *Processes* **2019**, *7*, 316. [CrossRef]
4. Algahtani, A.; Mahmoud, E.; Khan, S.; Tirth, V. Experimental Studies on Corrosion Behavior of Ceramic Surface Coating using Different Deposition Techniques on 6082-T6 Aluminum Alloy. *Processes* **2018**, *6*, 240. [CrossRef]
5. Nguyen, T.; Nguyen, M.; Nguyen, D.; Bach, L.; Lam, T. Model for Thin Layer Drying of Lemongrass (Cymbopogon citratus) by Hot Air. *Processes* **2019**, *7*, 21. [CrossRef]
6. Kern, T.; Krhač, E.; Senegačnik, M.; Urh, B. Digitalizing the Paints and Coatings Development Process. *Processes* **2019**, *7*, 539. [CrossRef]
7. Khan, A.; Nie, Y.; Shah, Z. Impact of Thermal Radiation and Heat Source/Sink on MHD Time-Dependent Thin-Film Flow of Oldroyed-B, Maxwell, and Jeffry Fluids over a Stretching Surface. *Processes* **2019**, *7*, 191. [CrossRef]
8. Khan, A.; Nie, Y.; Shah, Z. Impact of Thermal Radiation on Magnetohydrodynamic Unsteady Thin Film Flow of Sisko Fluid over a Stretching Surface. *Processes* **2019**, *7*, 369. [CrossRef]
9. Ullah, A.; Shah, Z.; Kumam, P.; Ayaz, M.; Islam, S.; Jameel, M. Viscoelastic MHD Nanofluid Thin Film Flow over an Unsteady Vertical Stretching Sheet with Entropy Generation. *Processes* **2019**, *7*, 262. [CrossRef]

Article

Fabrication of New Liquid Crystal Device Using Layer-by-Layer Thin Film Process

Gitae Moon [1,†], Wonjun Jang [2,†], Intae Son [1], Hyun A. Cho [2], Yong Tae Park [2,*] and Jun Hyup Lee [1,*]

[1] Department of Chemical Engineering, Myongji University, Yongin, Gyeonggi 17058, Korea; mgt1122@naver.com (G.M.); thsckdghgh@naver.com (I.S.)

[2] Department of Mechanical Engineering, Myongji University, Yongin, Gyeonggi 17058, Korea; zmdkskgo2@naver.com (W.J.); chocho9508@hanmail.net (H.A.C.)

* Correspondence: ytpark@mju.ac.kr (Y.T.P.); junhyuplee@mju.ac.kr (J.H.L.); Tel.: +82-31-330-6343 (Y.T.P.); +82-31-330-6384 (J.H.L.)

† These authors contributed equally to this work.

Received: 29 June 2018; Accepted: 26 July 2018; Published: 1 August 2018

Abstract: Indium tin oxide (ITO) transparent electrodes are troubled with high cost and poor mechanical stability. In this study, layer-by-layer (LBL)-processed thin films with single-walled carbon nanotubes (SWNTs) exhibited high transparency and electrical conductivity as a candidate for ITO replacement. The repetitive deposition of polycations and stabilized SWNTs with a negative surfactant exhibits sufficiently linear film growth and high optoelectronic performance to be used as transparent electrodes for vertically aligned (VA) liquid crystal display (LCD) cells. The LC molecules were uniformly aligned on the all of the prepared LBL electrodes. VA LCD cells with SWNT LBL electrodes exhibited voltage-transmittance (V-T) characteristics similar to those with the conventional ITO electrodes. Although the response speeds were slower than the LCD cell with the ITO electrode, as the SWNT layers increased, the display performance was closer to the LCD cells with conventional ITO electrode. This work demonstrated the good optoelectronic performance and alignment compatibility with LC molecules of the SWNT LBL assemblies, which are potential alternatives to ITO films as transparent electrodes for LCDs.

Keywords: electrode; layer-by-layer; liquid crystal device; thin film

---

1. Introduction

Liquid crystal displays (LCDs) are extensively utilized in daily life for items such as notebook computers, smartphones, and portable televisions due to their lightweight, low power consumption, and good mobility. Up to the present day, many studies have revealed several LC modes of display to achieve high picture qualities, e.g., in-plane switching (IPS) [1], pixel-patterned vertical alignment (VA) [2], and fringe-field switching (FFS) [3]. In an LCD, LC molecules located between two electrodes are moved by the applied electric field to transmit light, thereby displaying information.

Because of the growing interest in display applications, many studies on transparent electrodes have been recently conducted, focusing on electron transport and high transparency. In previous studies, metal oxides have been investigated as transparent electrodes because they display low resistance and high transparency [4,5]. Indium tin oxide (ITO) material has been commonly used for the last several decades as a transparent electrode thin film owing to its high visible transparency, excellent electrical conductivity, and availability of deposition on glass and plastic substrates. However, its high production cost and fragile characteristics limit its applicability for flexible electronic devices [6,7]. Also, the total reflection at the glass/ITO interface and the low adhesion to organic substrates lower the display performance of LC devices containing ITO transparent electrodes [8,9].

Many electrode materials, such as organic carbon nanotubes (CNTs) [10,11], metallic nanowires [12,13], metal thin films [14], graphene [15,16], and conductive polymers [17,18], have been studied as alternatives to ITO.

The single-walled carbon nanotube (SWNT) with a few nanometer radius is an emerging and especially promising material as alternative transparent electrodes. Several approaches have been utilized to generate SWNT films such as transfer printing [19], direct chemical vapor deposition (CVD) growth [20], vacuum filtration [21], and rod coating [22]. These SWNT thin films showed fairly high optical transmittance and good conducting properties which are similar to those of the conventional ITO-coated poly(ethylene terephthalate) (PET; 50–200 Ω resistance and ~83% transmission at 550 nm) substrate [23]. However, these methods tend to exhibit some problems such as poor film surface, optoelectronic performance, high cost, and a complex process [24]. Transfer printing, one of the most-used thin film techniques, results in difficult large-area fabrication and relatively brittle SWNT thin films. Therefore, in this study, we focused on the transparent electrode properties of SWNT by the layer-by-layer (LBL) process. Compared with other CNT thin film methods, the LBL technique provides the precise control needed to make thin films with a few nanometer thickness.

LBL assembly is one of the wet coating methods that has been widely studied in last 20 years due to its simple process and high controllability under ambient conditions [25,26]. This process can be used as a method of producing the thin films suitable for a variety of materials, which may include nanoparticles, nanotubes, nanowires, nanoplatelets, dyes, dendrimers, proteins, and viruses. The LBL process enables the production of homogeneous multifunctional multilayers with a controlled method. Moreover, it can be applied to a variety of substrates (e.g., glass and quartz slides, silicon wafers, and polymeric films) to evaluate the performance of electrical conductivity [27,28], sensing [29,30], and energy harvesting [31].

In this study, VA LCD devices with a polyimide (PI) alignment layer on SWNTs and poly (diallyldimethylammonium chloride) (PDDA) layer-by-layer electrodes were fabricated and characterized to evaluate their potential for transparent organic electrode materials in an LC display. Firstly, PDDA was used as the positively charged polyelectrolyte in conjunction with a negatively charged surfactant to deposit SWNTs onto a PDDA layer. The alternate dipping in positively and negatively charged solutions resulted in SWNT/polymer thin film deposition onto the substrate. The LC test devices were fabricated by filling the LCs into the inner parts of the PI layer by spin coating, and their device structures have been reported in the literature [32,33]. The electro-optical (E-O) properties of the prepared VA LCD devices containing the LBL transparent electrodes were examined and compared to those of LCD devices with conventional ITO electrodes.

2. Materials and Methods

2.1. Materials

Purified electric arc SWNTs (TUBALL™ SWNT, individual tube: average 1 μm length and 2 nm diameter, C > 75%) were used in this study. PDDA (M_w ~200,000–350,000 g/mol), sodium deoxycholate (DOC, $C_{24}H_{39}NaO_4$), isopropyl alcohol (IPA), and methanol were purchased from Sigma-Aldrich (Yongin, Korea). Nematic LCs ($\Delta n = 0.095$, $\Delta\varepsilon = -3.1$), PI (AL607XX, JSR, Tokyo, Japan), ITO glass (SNT 15Ω, FINE CHEMICAL, Seoul, Korea), UV-curable sealant (SD-2X, Sekisui Chemical, Tokyo, Japan), and ball spacers with a diameter of 3.25 μm (SP-205XX, Sekisui Chemical, Tokyo, Japan) were used to fabricate the LCD. All chemicals were used as received. Slide glasses, PET films (100 μm thickness, Goodfellow, Huntingdon, England), and quartz crystals with 5 MHz gold-electrodes (Maxtek, Inficon Korea, Seongnam, Korea) were purchased for analysis of UV–Visible (UV–Vis) light absorbance, field emission scanning electron microscopy (FE-SEM), and mass, respectively.

2.2. Fabrication of LCD Cells Using SWNT LBL Transparent Electrodes

Both positively and negatively charged aqueous solutions were obtained by using 18.2 MΩ deionized (DI) water (YL aquaMAX Ultra 370, YL Instrument, Anyang, Korea). A PDDA solution with a concentration of 0.25 wt. % as a cationic solution was prepared. An SWNT dispersion (0.05 wt. %) treated by 1.5 wt. % DOC in DI water (SWNT–DOC dispersion) was prepared as a negatively charged solution. The large SWNT bundles in the SWNT–DOC dispersion were removed by tip ultrasonication (HD-1070, Sonopuls, Berlin, Germany) for 45 min. Then, each surface of the slide glasses was cleaned by a plasma treatment (PDC 32G-2, Harrick Plasma, Ithaca, NY, USA) for 10 min. The PET films were cleaned with IPA, methanol, and DI water, dried thoroughly with clean air, and then also treated by the plasma etcher for 5 min. For LBL deposition, PET films and glass slides were alternately dipped into PDDA and SWNT–DOC solutions with rinsing and drying, beginning with the PDDA. Each deposition cycle resulted in one bilayer (BL) on the substrate. For the cyclic deposition, the PDDA and SWNT–DOC layers were formed by 5 min of dipping for each, followed by rinsing and drying. The schematic representation of the LBL thin film process is shown in Figure 1a. The notation of [PDDA/(SWNT–DOC)]$_n$ (n = the number of BLs on the substrate) represents the deposition cycle of LBL thin films. Light absorbance was measured between 400 and 800 nm and transmittance was calculated from the absorbance measured by a UV–Vis spectrometer (USB2000, Ocean Optics, Wonwoo Systems, Seoul, Korea). The mass growth from the LBL deposition was measured by a quartz crystal microbalance (QCM, QCM 200, Standard Research System, Inficon Korea, Seongnam, Korea). Sheet resistance was measured with a digital multimeter (Model2000, Keithley, Seoul, Korea). Surface images of LBL thin films were acquired with a FE-SEM (SU-70, Hitachi High-Technologies, Seoul, Korea) at the voltage of 15 kV.

(a)

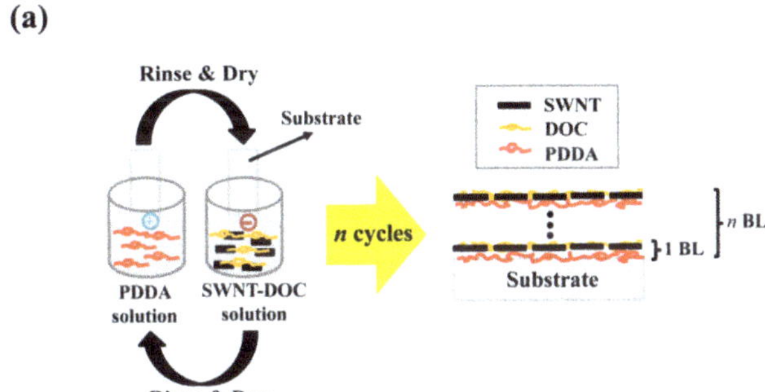

Figure 1. *Cont.*

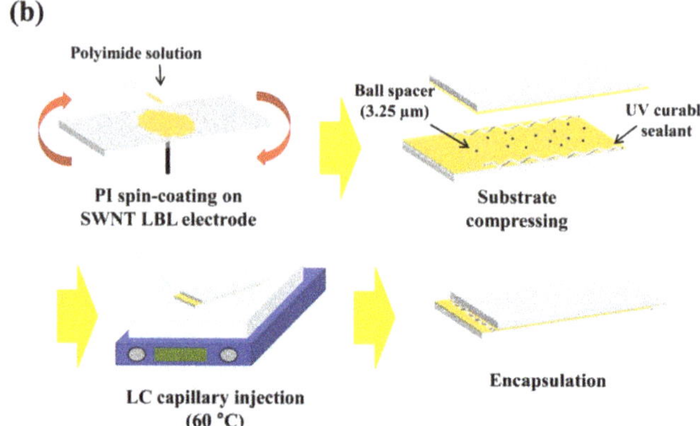

Figure 1. (a) Schematic illustration of the layer-by-layer (LBL) dipping process with poly (diallyldimethylammonium chloride) (PDDA) and single-walled carbon nanotube–sodium deoxycholate (SWNT–DOC) aqueous solutions. (b) Manufacturing process of liquid crystal display (LCD) using transparent SWNT electrode fabricated by the LBL method.

The overall process for manufacturing the LCD using the transparent SWNT-based electrode fabricated by the LBL method is shown in Figure 1b. PI was uniformly coated on the LBL film at 3500 rpm for 30 s to form alignment layers. These PI layers on the LBL film were imidized at 230 °C for 30 min. The LBL-LCD cell was manufactured by coupling the two LBL substrates. The cell gap of the LBL-LCD device was formed by using 3.25-µm ball spacers. The ball spacers were dispersed in an ethanol solution and then sprayed on the PI-coated LBL electrode. An empty cell for LC injection was prepared by applying the UV-curable sealant on the PI-coated substrate and then attaching the upper and lower substrates together. The sealant was cured by exposure to UV radiation of 3.0 J·cm^{-2} (WUV-L50, DAIHAN, Seoul, Korea). After the empty LBL-LCD cell was filled with nematic LCs at 60 °C using capillary force, the inlet and outlet areas were sealed using the UV-curable sealant. The manufactured LBL-LCD cell was thermally treated for 1 h at 100 °C. The LC alignment was examined by a polarized optical microscope (POM; BX51, OLYMPUS, Tokyo, Japan) with a crossed polarizer. To evaluate the E-O characteristics, the voltage-transmittance (V-T) curve and on/off response time of the LCD cell were investigated. Further, to investigate the E-O properties according to the number of bilayers, the PI layer was spin coated on 2, 6, and 10 BLs electrodes to fabricate LCD cells with different configurations.

3. Results and Discussion

Figure 2 shows the image of 2–10 BLs SWNT LBL thin films deposited on the glass substrates, the SEM surface image of these PDDA/SWNT LBL assemblies on the PET substrate, and a photograph of the prepared LC cell using an SWNT LBL electrode. The growth of SWNT deposition was observed as the increase in the number of BLs deposited and was confirmed using QCM mass and UV–Vis absorbance measurements, as shown in Figure 3. In a SEM image of Figure 2b, the LBL thin film with 10 BLs exhibits a good distribution of SWNTs owing to the exfoliation of individual SWNTs in water by DOC [27,28]. The dispersion of SWNTs provides the contacts within the nanotube network, indicating a transport route of electrons through the film. Also, SWNT LBL deposition was quite uniform. SWNT–DOCs are uniformly deposited onto the PDDA layer and the following PDDA deposits are thick enough to lead to the smooth surface of thin film.

Figure 3 shows the absorbance spectra of the SWNT LBL thin films from 1 to 10 BLs at wavelengths between 450 and 800 nm and the mass growth curve of the LBL thin films corresponding to the number of bilayers deposited. Because the absorbance of UV–Vis light increases in proportion to the thickness of thin films, Figure 3a exhibits the almost linear growth of LBL multilayers (i.e., PDDA and SWNT–DOC BLs) up to 10 BLs. Moreover, the Table 1 shows that the transparency decreases with increasing BLs in the visible light wavelength, suggesting that the SWNT is coated proportionally as BLs are deposited. As the number of BLs increases, the electrical conductivity decreases because the amount of SWNTs coated on the glass slide increases proportionally. Similar to the UV–Vis absorbance, the mass of [PDDA/(SWNT–DOC)]$_n$ measured by QCM in Figure 3b shows a quite linear increase with the number of BLs, suggesting constant deposition up to 10 BLs.

Figure 4 is a graph showing the optoelectronic relationship of the LBL thin films by showing electric resistance as a function of optical transparency. Both resistance and transparency decreased as the number of BLs increased. In the previous studies, the thicker SWNT films showed the higher conductivity due to a more three-dimensional SWNT network [27,28]. For the reason of more efficient electron transport, the resistance of LBL electrodes decreases according to the number of BLs. The amount of SWNTs coated on the substrate increases along with the number of BLs, so the optical transparency decreases. The electrical and optical properties of some SWNT LBL electrodes used for LCD cells are also summarized and compared with ITO in Table 1. Despite the relatively low resistance of 1 kΩ at high BLs, it should be even lower for a reasonable ITO replacement.

Table 1. Electric and optical properties of ITO and SWNT LBL electrodes.

Electrodes	Resistance (kΩ)	Transparency (at 550 nm)
ITO	0.02	84.14
SWNT 2 BLs	29	95.50
SWNT 6 BLs	1.3	82.22
SWNT 10 BLs	0.5	69.02

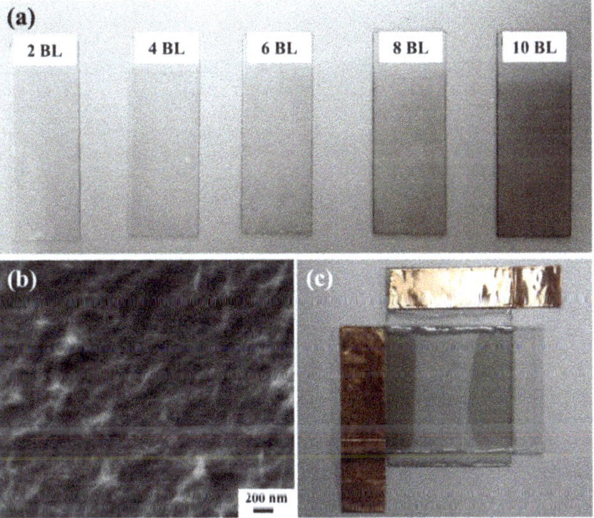

Figure 2. (a) Photograph of [PDDA/(SWNT–DOC)]$_n$ (n = 2–10) assemblies on both sides of a glass slide. (b) SEM surface image of [PDDA/(SWNT–DOC)]$_{10}$ on a poly(ethylene terephthalate) (PET) film. (c) Photograph of an LC cell using an SWNT LBL electrode.

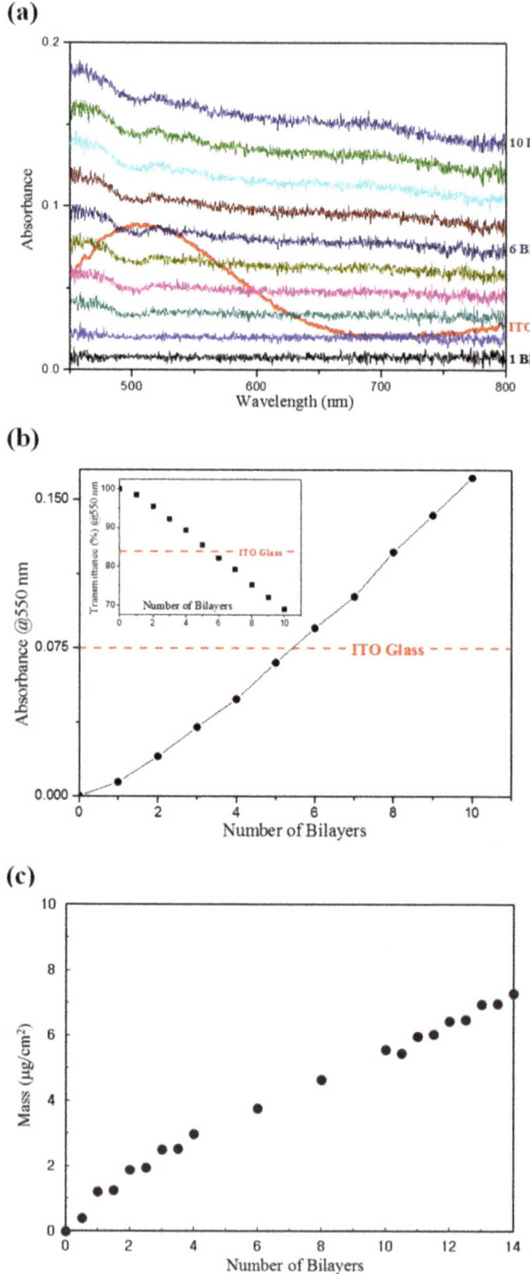

Figure 3. (**a**) UV–Visible (UV–Vis) absorbance spectra of [PDDA/(SWNT–DOC)]$_n$ (n = 1–10) thin films and an indium tin oxide (ITO) coating from 450 to 800 nm wavelength. (**b**) Light absorbance and transmittance (inset) of 1–10 bilayers (BLs) at 550 nm wavelength. (**c**) Mass growth of a [PDDA/(SWNT–DOC)] system up to 14 BLs (half BL:PDDA deposition).

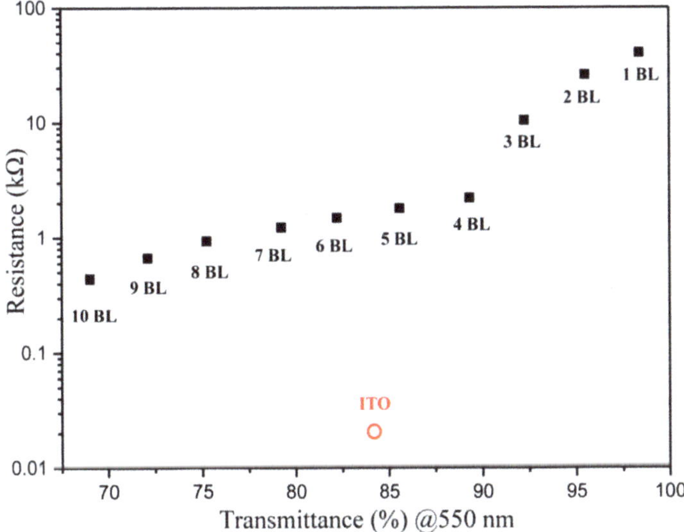

Figure 4. Optoelectronic performance (electrical resistance vs. optical transmittance at 550 nm wavelength) of [PDDA/(SWNT–DOC)]$_n$ thin films as a function of the number of bilayers deposited to 10 BLs.

The POM photographs of the VA LCDs with the crossed polarizer set in two orientations with respect to the LCD cells are shown in Figure 5. Figure 5a exhibits the image of the LCD cell with the polarizer in the same direction as the cell. The LCD cell using the LBL electrode exhibited almost the same image as the LCD cell using the ITO substrate. However, as the number of BLs increased, the LCD cell showed a very fine texture. The LCD cell using the 2 BL SWNT electrode showed no heterogeneity, and the LCD cell using the 10 BL electrode showed a fine texture, probably due to impurities incorporated in the LBL process. Figure 5b shows the image of the LCD cell under cross polarization. The LCD cell using the LBL electrode exhibited a nearly uniform black image, similar to that exhibited by the cell using the ITO substrate. Moreover, the slight heterogeneity in polarizers in the same direction was not observed at all in the black images. Under cross polarization, the VA LCD cells with LBL electrodes showed lower transmittances than the cell with ITO electrodes. While VA LCD cells with ITO showed a transmittance of 0.084 (100%), VA LCD cells with LBL electrodes exhibited 0.077 (91.7%), 0.059 (70.2%), and 0.043 (51.2%) for 2 BL, 6 BL, and 10 BL, respectively. These results indicated the creation of uniform homeotropic LC alignment on the LBL thin film electrode; the number of BLs did not disturb the vertical alignment of the nematic LCs in the device. In addition, typical conoscopic images of the VA LCDs are shown in the insets of Figure 5 [34,35]. The conoscopic images showed a distinct Maltese cross, which indicates the vertical alignment of the LC molecules.

To confirm the effect of the number of BLs, we measured the V-T characteristics and response time properties of the newly manufactured LCDs, which represented their E-O properties. The voltage-dependent optical transmittance of the VA LCDs is shown in Figure 6. The shapes of the V-T curves of VA LCD cells using the SWNT LBL electrodes in the transmittance range of 10% 90% were similar to those of conventional VA LCD cells using the ITO electrode. In the V-T curves, threshold voltages at a transmittance of 10% were 3.34, 2.38, 2.34, and 2.23 V for cells with the 2-BL, 6-BL, 10-BL, and ITO electrodes, respectively. It is noteworthy that the threshold voltage of the VA LCD cells containing the LBL electrode was higher than that of the LC device with the conventional ITO electrode. This means that the LC molecules form a stable vertical alignment similar to that using the ITO electrode. However, the 2-BL electrode showed the highest threshold voltage of 3.34 V.

This result suggests that the high resistance due to the small SWNT layers causes the LC molecules to drive at a high voltage. As the number of BLs increased, the electrical resistance decreased, and the threshold voltage tended towards values closer to that of the LCD cell using the conventional ITO electrode. Particularly noteworthy is the fact that the driving voltage of LCD cells with 6 BLs or 10 BLs is lower than that of LCD cells with ITO electrodes. While the LC device with ITO showed a white state at 2.8 V, the LCD containing 6 BLs or 10 BLs exhibited an on-state at 2.4 V. As a result, it was confirmed that there is no problem in driving the LC device with 6 BLs or more SWNT layers.

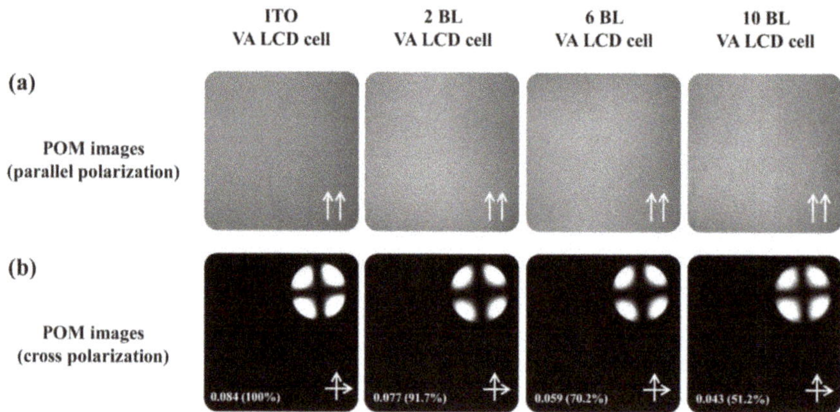

Figure 5. Vertically aligned (VA) LCDs observed by a polarized optical microscope (POM) with crossed polarizers set in two orientations with respect to the LCD cells: (**a**) parallel polarization and (**b**) cross polarization. The inserted values are the transmittances of VA LCD cells under cross polarization using a UV–Vis spectrometer.

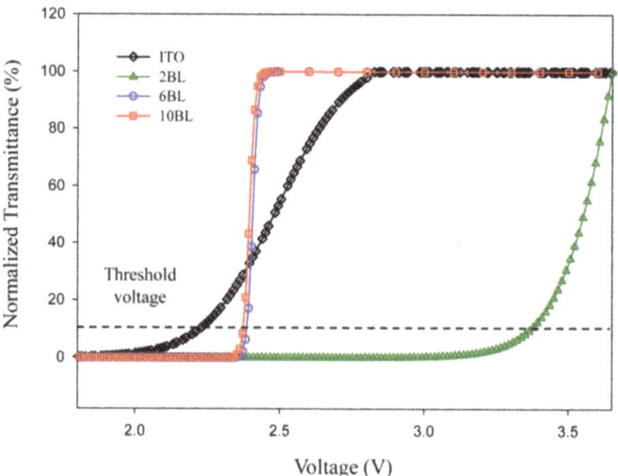

Figure 6. Voltage-transmittance (V-T) curves of VA LCD cells using 2-BL, 6-BL, 10-BL, and ITO electrodes.

We confirmed the switching drive of the LCDs by measuring the response time of the newly manufactured LCD cell. The response time characteristics were confirmed by measuring the transmittance caused by the alignment change of the LC molecules by applying a square wave of 120 Hz to the LCD cell using a waveform generator. Figure 7a exhibits the rising time curves of the

LCDs fabricated with the ITO and SWNT LBL electrodes, and Figure 7b shows their falling time curves. The LCD cell using the LBL electrode could be switched on or off by applying voltage, similar to the cell using the conventional ITO substrate. However, due to the high electrical resistance of the LBL thin film, a time delay occurred until the square wave was actually applied. Due to this time delay, the response characteristics of the LCD cell using the LBL electrode was slower than that of the LC device with the conventional ITO electrode. The VA LCD cells manufactured using the conventional ITO electrode had a rising time of 22 ms and a falling time of 15 ms. The VA LCD cell using the 10-BL electrode had on and off times of 100 ms and 106 ms, respectively, because it had the lowest electrical resistance among the LBL electrodes. The VA LCD cell using the 6-BL electrode had rising and fall times of 139 ms and 152 ms, respectively. The fabricated VA LCD cell using the 2-BL electrode had a 201 ms on time and a 209 ms off time. Thus, it was confirmed that the response time of the prepared LC device using the LBL electrode was slower than that using the conventional ITO electrode, but the response time became closer to that of the ITO electrode as the number of BLs increased.

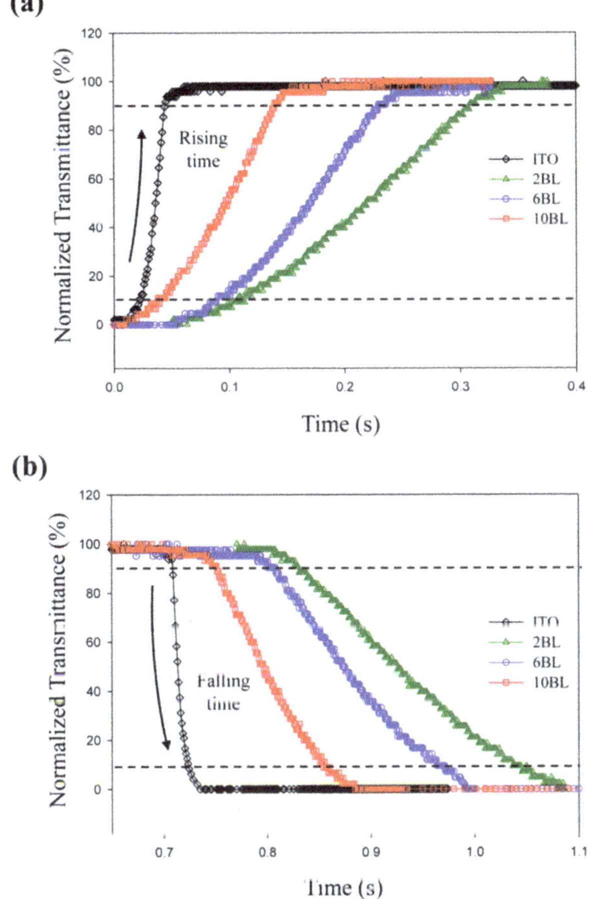

Figure 7. Response time curves of VA LCD cells using 2-BL, 6-BL, 10-BL, and ITO electrodes: (**a**) rising time curves and (**b**) falling time curves.

4. Conclusions

In this study, transparent electrodes for LCDs were fabricated with SWNT thin films by the LBL method to replace conventional ITO electrodes. SWNT LBL films with lower than 1.3 kΩ of electrical resistance and greater than 85% of optical transparency were fabricated by repetitive immersion of the substrate into positively and negatively charged polyelectrolyte solutions. The nematic LCs were uniformly aligned by spin coating a VA PI film on the prepared LBL electrode. The POM images of the VA LCD cell using the LBL electrode confirmed that the LC molecules were uniformly aligned. Moreover, the number of bilayers did not influence the homeotropic alignment of the nematic LC materials. Also, LCD cells using the proposed electrode exhibited V-T characteristics similar to that of VA LCD cells using the conventional ITO electrode, with its on/off state depending on the application of voltage. Due to the high electrical resistance of the LBL electrode, the response speed characteristics of the LCD device using the LBL electrode were slower than that of the LC display using the conventional ITO electrode. However, as the number of BLs increased, the threshold voltage and response time tended to be closer to that of the ITO electrode and the response speed became faster. These results demonstrated the potential of the LBL thin films as good alternatives to ITO films as transparent conducting electrodes for LCDs. The combination of simple processing and good optoelectronic performance should make SWNT LBL thin films very promising for future display applications.

Author Contributions: G.M., Y.T.P., and J.H.L. conceptualized the phenomenon. Y.T.P. and J.H.L. designed the project. G.M., W.J., I.S., and H.A.C. conducted the experiments. All authors wrote the manuscript and analyzed the data. Y.T.P. and J.H.L. supervised the whole project.

Funding: This work was financially supported by the Mid-career Researcher Program (No. 2017R1A2B4006104) through a National Research Foundation of Korea (NRF) grant funded by the Ministry of Science, ICT and Future Planning (MSIP). This work was also supported by the National Research Foundation of Korea (NRF) grant funded by the Korea government (MSIT) (No. NRF-2018R1A5A1024127).

Conflicts of Interest: The authors declare no conflict of interest. The funders had no role in the design of the study; in the collection, analyses, or interpretation of data; in the writing of the manuscript, and in the decision to publish the results.

Nomenclature

BL	Bilayer
CNT	Carbon nanotube
CVD	Chemical vapor deposition
DI	Deionized, DOC: Sodium deoxycholate
E-O	Electro-optical
FE-SEM	Field emission scanning electron microscopy
FFS	Fringe-field switching
IPA	Isopropylalcohol
IPS	In-plane switching
ITO	Indium tin oxide
LBL	layer-by-layer
LCD	Liquid crystal display
PDDA	Poly(diallyldimethylammonium chloride)
PET	Poly(ethylene terephthalate)
PI	Polyimide
POM	Polarized optical microscope
QCM	Quartz crystal microbalance
SWNT	Single-walled carbon nanotube
T_{NI}	Nematic-to-isotropic transition temperature
UV–Vis	UV–Visible spectroscopy
VA	Vertical alignment
V-T	Voltage-transmittance

References

1. Ohe, M.H.; Kondo, K.S. Electro-optical characteristics and switching behavior of the in-plane switching mode. *Appl. Phys. Lett.* **1995**, *67*, 3895–3897. [CrossRef]
2. Lee, Y.J.; Kim, Y.K.; Jo, S.I.; Bae, K.S.; Choi, B.D.; Kim, J.H.; Yu, C.J. Fast vertical alignment mode with continuous multi-domains for a liquid crystal display. *Opt. Express* **2009**, *17*, 23417–23422. [CrossRef] [PubMed]
3. Lee, S.H.; Lee, S.L.; Kim, H.Y. Electro-optic characteristics and switching principle of a nematic liquid crystal cell controlled by fringe-field switching. *Appl. Phys.* **1998**, *73*, 2881–2883. [CrossRef]
4. Igasaki, Y.; Saito, H. Substrate temperature dependence of electrical properties of ZnO:Al epitaxial films on sapphire. *J. Appl. Phys.* **1991**, *69*, 2190–2195. [CrossRef]
5. Lan, Y.F.; Peng, W.C.; Lo, Y.H.; He, J.L. Indium tin oxide films deposited by thermionic-enhanced DC magnetron sputtering on unheated polyethylene terephthalate polymer substrate. *Mater. Res. Bull.* **2009**, *44*, 1760–1764. [CrossRef]
6. Lin, Y.C.; Shi, W.Q.; Chen, Z.Z. Effect of deflection on the mechanical and optoelectronic properties of indium tin oxide films deposited on polyethylene terephthalate substrates by pulse magnetron sputtering. *Thin Solid Films* **2009**, *517*, 1701–1705. [CrossRef]
7. Gu, G.; Shen, Z.; Burrows, P.E.; Forrest, S.R. Transparent flexible organic light-emitting devices. *Adv. Mater.* **1997**, *9*, 725–728. [CrossRef]
8. Cairns, D.R.; Witte, R.P.; Sparacin, D.K.; Sachsman, S.M.; Paine, D.C.; Crawford, G.P.; Newton, R.P. Strain-dependent electrical resistance of tin-doped indium oxide on polymer substrates. *Appl. Phys. Lett.* **2000**, *76*, 1425–1427. [CrossRef]
9. Tao, C.S.; Jiang, J.; Tao, M. Natural resource limitations to terawatt-scale solar cells. *Sol. Energy Mater. Sol. Cells* **2011**, *95*, 3176–3180. [CrossRef]
10. Wu, Z.; Chen, Z.; Du, X.; Logan, J.M.; Sippel, J.; Nikolou, M.; Kamaras, K.; Reynolds, J.R.; Tanner, D.B.; Hebard, A.F.; et al. Transparent, conductive carbon nanotube films. *Science* **2004**, *305*, 1273–1276. [CrossRef] [PubMed]
11. Zhang, M.; Fang, S.; Zakhidov, A.A.; Lee, S.B.; Aliev, A.E.; Williams, C.D.; Atkinson, K.R.; Baughman, R.H. Strong, transparent, multifunctional, carbon nanotube sheets. *Science* **2005**, *309*, 1215–1219. [CrossRef] [PubMed]
12. Lee, J.Y.; Connor, S.T.; Cui, Y.; Peumans, P. Solution-Processed Metal Nanowire Mesh Transparent Electrodes. *Nano Lett.* **2008**, *8*, 689–692. [CrossRef] [PubMed]
13. Hsu, P.C.; Wang, S.; Wu, H.; Narasimhan, V.K.; Kong, D.; Lee, H.R.; Cui, Y. Performance enhancement of metal nanowire transparent conducting electrodes by mesoscale metal wires. *Nat. Commun.* **2013**, *25*, 2522–2528. [CrossRef] [PubMed]
14. Meiss, J.; Riede, M.K.; Leo, K. Optimizing the morphology of metal multilayer films for indium tin oxide (ITO)-free inverted organic solar cells. *Appl. Phys. Lett.* **2009**, *105*, 063108-1–063108-5. [CrossRef]
15. Tung, V.C.; Chen, L.M.; Allen, M.J.; Wassei, J.K.; Nelson, K.; Kaner, R.B.; Yang, Y. Low-temperature solution processing of graphene-carbon nanotube hybrid materials for high-performance transparent conductors. *Nano Lett.* **2009**, *9*, 1949–1955. [CrossRef] [PubMed]
16. Mei, X.G.; Ouyang, J.Y. Ultrasonication-assisted ultrafast reduction of graphene oxide by zinc powder at room temperature. *Carbon* **2011**, *49*, 5389–5397. [CrossRef]
17. Alemu, D.L.; Wei, H.Y.; Ho, K.C.; Chu, C.W. Highly conductive PEDOT:PSS electrode by simple film treatment with methanol for ITO-free polymer solar cells. *Energy Environ. Sci.* **2012**, *5*, 9662–9671. [CrossRef]
18. Na, S.I.; Kim, S.S.; Jo, J.; Kim, D.Y. Efficient and Flexible ITO-Free Organic Solar Cells Using Highly Conductive Polymer Anodes. *Adv. Mater.* **2008**, *21*, 4061–4067. [CrossRef]
19. Zhang, D.; Ryu, K.; Liu, X.; Polikarpov, E.; Ly, J.; Tompson, M.E.; Zhou, C. Transparent, conductive, and flexible carbon nanotube films and their application in organic light-emitting diodes. *Nano Lett.* **2006**, *6*, 1880–1886. [CrossRef] [PubMed]
20. Ma, W.; Song, L.; Yang, R.; Zhang, T.; Zhao, Y.; Sun, L.; Zhang, Z. Directly synthesized strong, highly conducting, transparent single-walled carbon nanotube films. *Nano Lett.* **2007**, *7*, 2307–2311. [CrossRef] [PubMed]

21. Green, A.A.; Hersam, M.C. Colored Semitransparent Conductive Coatings Consisting of Monodisperse Metallic Single-Walled Carbon Nanotubes. *Nano Lett.* **2008**, *8*, 1417–1422. [CrossRef] [PubMed]
22. Dan, B.; Irvin, G.C.; Pasquali, M. Continuous and scalable fabrication of transparent conducting carbon nanotube films. *ACS Nano* **2009**, *3*, 835–843. [CrossRef] [PubMed]
23. Park, Y.T.; Ham, A.Y.; Yang, Y.H.; Grunlan, J.C. Fully organic ITO replacement through acid doping of double-walled carbon nanotube thin film assemblies. *RSC Adv.* **2011**, *1*, 662–671. [CrossRef]
24. Yu, X.; Rajamani, R.; Stelson, K.A.; Cui, T. Fabrication of carbon nanotube based transparent conductive thin films using layer-by-layer technology. *Surf. Coat. Technol.* **2008**, *202*, 2002–2007. [CrossRef]
25. Decher, G.; Lvov, Y.; Schmitt, J. Proof of multilayer structural organization in self-assembled polycation-polyanion molecular films. *Thin Solid Films* **1994**, *244*, 772–777. [CrossRef]
26. Hammond, P.T. Form and function in multilayer assembly: New applications at the nanoscale. *Adv. Mater.* **2004**, *16*, 1271–1293. [CrossRef]
27. Park, Y.T.; Ham, A.Y.; Grunlan, J.C. High electrical conductivity and transparency in deoxycholate-stabilized carbon nanotube thin films. *J. Phys. Chem. C* **2010**, *114*, 6325–6333. [CrossRef]
28. Park, Y.T.; Ham, A.Y.; Grunlan, J.C. Heating and acid doping thin film carbon nanotube assemblies for high transparency and low sheet resistance. *J. Mater. Chem.* **2011**, *21*, 363–368. [CrossRef]
29. Park, J.J.; Hyun, W.J.; Mun, S.C.; Park, Y.T.; Park, O.O. Highly stretchable and wearable graphene strain sensors with controllable sensitivity for human motion monitoring. *ACS Appl. Mater. Interfaces* **2015**, *7*, 6317–6324. [CrossRef] [PubMed]
30. Zhou, M.; Dong, S. Bioelectrochemical interface engineering: Toward the fabrication of electrochemical biosensors, biofuel cells, and self-powered logic biosensors. *Accounts Chem. Res.* **2011**, *44*, 1232–1243. [CrossRef] [PubMed]
31. Chung, I.J.; Kim, W.; Jang, W.; Park, H.W.; Sohn, A.; Chung, K.B.; Kim, D.W.; Choi, D.; Park, Y.T. Layer-by-layer assembled graphene multilayers on multidimensional surfaces for highly durable, scalable, and wearable triboelectric nanogenerators. *J. Mater. Chem. A* **2018**, *6*, 3108–3115. [CrossRef]
32. Wang, H.Y.; Wu, T.X.; Zhu, X.Y.; Wu, S.T. Correlations between liquid crystal director reorientation and optical response time of a homeotropic cell. *J. Appl. Phys.* **2004**, *95*, 5502–5508. [CrossRef]
33. Seo, D.S.; Kim, J.H. Generation of pretilt angle in NLC and EO characteristics of transcription-aligned TNLCD fabricated by transcription alignment on polyimide surfaces. *Liq. Cryst.* **1999**, *26*, 397–400. [CrossRef]
34. Kim, T.H.; Ju, C.H.; Kang, H. Vertical alignment of liquid crystal on tocopherol substituted polystyrene films. *Liq. Cryst.* **2018**, *45*, 801–810. [CrossRef]
35. Oh, S.K.; Nakagawa, M.S.; Ichimura, K.H. Relationship between the ability to control liquid crystal alignment and wetting properties of calix[4]resorcinarene monolayers. *J. Mater. Chem.* **2001**, *11*, 1563–1569. [CrossRef]

Article

Comparison of Growth Characteristics and Properties of CVD TiN and TiO$_2$ Anti-Coking Coatings

Shiyun Tang [1,2,*], Tao Liu [1], Shuiping Duan [1], Junjiang Guo [1,2] and Anjiang Tang [1,2,*]

1   College of Chemical Engineering, Guizhou Institute of Technology, Guiyang 550003, China
2   Engineering Technology Research Center of Fluorine and Silicon Material, Guiyang 550003, China
*   Correspondence: 20150649@git.edu.cn (S.T.); 20140393@git.edu.cn (A.T.); Tel.: +86-0851-8834-9137 (S.T.)

Received: 20 July 2019; Accepted: 26 August 2019; Published: 30 August 2019

Abstract: Coating metals with anti-coking materials inhibit their catalytic coking and are especially beneficial in the pyrolysis of hydrocarbon fuels. It is believed that growth characteristics and properties may play a pivotal role in the anti-coking performance of chemical vapor deposition (CVD) coatings. In this study, TiN and TiO$_2$ coatings were obtained by CVD using TiCl$_4$–N$_2$–H$_2$ and TiCl$_4$–H$_2$–CO$_2$ systems, respectively. The effects of deposition time, residence time, and partial pressure were examined, and the coating microstructure was characterized by scanning electron microscopy (SEM). The results reveal that the effect of deposition parameters on the growth characteristics of TiN and TiO$_2$ coatings is very different. The growth of the TiN coating shows characteristics of the island growth model, while the TiO$_2$ coating follows the layer model. In general, the growth rate of the star-shaped TiN crystals is higher than that of crystals of other shapes. For the TiO$_2$ coating, the layer mode growth characteristics indicate that the morphology of the TiO$_2$ coating does not change significantly with the experimental conditions. Coking tests showed that the morphology of non-catalytic cokes is not only affected by the temperature, pressure, and coking precursor, but is also closely related to the surface state of the coatings. Both TiN and TiO$_2$ coatings can effectively prevent catalytic coking and eliminate filamentous cokes. In some cases, however, the N or O atoms in the TiN and TiO$_2$ coatings may affect common carbon deposits formed by non-catalytic coking, such as formation of needle-like and flaky carbon deposits.

Keywords: hydrocarbon fuel; coke; anti-coking coating; growth characteristics

---

1. Introduction

The formation of cokes on metallic surfaces from thermal decomposition of hydrocarbon fuels pyrolysis is a major concern in the development of hypersonic aircraft, in which the fuel serves as not only the propellant, but also the ideal coolants to resolve the problem of thermal management by removing the waste heat from aircrafts with the physical and chemical heat sink (sensible heat-absorbing and cracking heat-absorbing) [1–3]. Not only that, coking and anti-coking have become the important subject of attention for many researchers at home and abroad, in the chemical, energy, environmental protection, aerospace, military, and other fields [4–6]. The formation of cokes is often accompanied by metal dusting, which is a serious form of carburization of Fe-, Ni-, and Co-base alloys. It not only leads to the disintegration of the metal substrate into small metal particles, leaving pits and grooves, but also causes severe catalytic coking in its wake, thus deteriorating the heat transfer process by sharply increasing heat resistances and weakening the substrate [7]. Recently, the coating method has been used to inhibit the catalytic coking of metal, and this has been especially beneficial during the pyrolysis of hydrocarbon fuels. Eser et al. [8] studied alumina, zirconia (ZrO$_2$), tantalum oxide (Ta$_2$O$_5$), and platinum coatings obtained by metal organic chemical vapor deposition (MOCVD) on AISI 304 foils to alleviate coking due to thermal oxidation in the thermal stressing of Jet A (350 °C and

3.5 MPa, 1 mL/min for 5 h), and observed that the effectiveness of the coatings in mitigating carbon deposition decreases in the following order: Pt > Ta_2O_5 > alumina from acetylacetonate > ZrO_2 > alumina from aluminum trisecondary butoxide > AISI 304. Similarly, Liu et al. [9] investigated a series of alumina coatings with different thicknesses (318–1280 nm) in SS321 tubes (2 mm i.d.) obtained by MOCVD using aluminum trisecondary butoxide and evaluated their anti-coking performance during the thermal cracking of Chinese RP-3 jet fuel under supercritical conditions (inlet temperature, 575 °C; outlet temperature, 650 °C; pressure, 5 MPa). Their results indicated that the anti-coking performance increased from 37% to 69% as the thickness of alumina coatings increased from 318 to 1280 nm. In addition, Xu et al. [10–12] published a series of articles on the anti-coking property of the SiO_2/S composite coating during light naphtha steam cracking and reported that the SiO_2/S coating reduces coke yield by 60% compared with the uncoated tube. More recently, Liu et al. [13] prepared a $MnCr_2O_4$ spinel coating on HP40 alloy by pack cementation and subsequent thermal oxidation and applied it to inhibit coke formation during light naphtha thermal cracking. Their result showed that the anti-coking performance of the $MnCr_2O_4$ spinel coating exceeded 60% during the thermal cracking of light naphtha. In previous works, the anti-coking performances of TiN, TiO_2, and TiC coatings were investigated [14–17]. The results show that TiN, TiO_2, and TiC coatings have an obvious anti-coking effect; especially, the anti-coking performance of the TiN coating exceed 90%, and the anti-coking effect of the coatings vary in the order TiN = TiC > TiO_2. Therefore, the anti-coking performance of Ti-based coatings obtained by CVD is often superior to that of the common SiO_2, Al_2O_3, and $MnCr_2O_4$ spinel coatings.

Although the TiN and TiC coatings show excellent anti-coking properties, the nature and mechanism of their anti-coking effect is not very clear. It is well known that coke is mainly formed on the interface of the metal substrate with hydrocarbon fuels, and that both the chemical composition and surface roughness affect the coking process significantly. With regard to the chemical composition, many studies have shown that some metals, such as Fe, Ni, and Mo, on the surface of the reaction channels have a catalytic effect for coking, while other metals, such as Cu, Si, Al, Cr, Ti, Nb, and Ta, are found to inhibit coking [18–20]. With regard to the surface roughness, Crynes et al. [21] examined the effect of surface roughness on coke formation during the pyrolysis of light hydrocarbon feedstock using polished metal coupons of Incoloy 800. The results showed that the coking rate reduced sharply after polishing; the ratio of carbon formed on the unpolished coupons to the polished ones varied from 5.6 for iso-butane to 28.1 for ethene. Similarly, Marek and Albright [22,23] and Gregg and Leach [24] also found reduced coke formation on the metal surface after polishing. However, Durbin and Castle [25] found that the amount of coke on a polished specimen did not drop compared with the unpolished specimen during acetone pyrolysis. They explained that this might be because of the increasing temperature on the metal surface during the process of heat treatment. In conclusion, it is believed that the growth characteristics and properties may play a pivotal role in the anti-coking performance of CVD coatings.

Furthermore, on the basis of previous works, the TiN coating has excellent anti-coking performance, but poor oxidation resistance, while the TiO_2 coating has good oxidation resistance at high temperature, but poor anti-coking performance [26]. This paper tries to provide more information on the relationship between the anti-coking performance and the growth characteristics and properties of CVD TiN and TiO_2 anti-coking coatings, and thus provide guidance for designing better anti-coking coatings.

2. Experimental Details

The preparation of TiN and TiO_2 coatings was conducted in a horizontal and tubular hot-wall reactor with 11 mm i.d. and 300 mm in length under atmospheric pressure. The 310S stainless steel foils (10 mm × 10 mm × 0.9 mm) were selected as the substrate. In general, preparation of TiN coating by the CVD method includes the following two systems: the $TiCl_4$–NH_3–H_2 system and the $TiCl_4$–N_2–H_2 system. Their basic chemical equations for the preparation of the TiN coating are as follows:

$$2TiCl_4 \text{ (g)} + 2NH_3 \text{ (g)} + H_2 \rightarrow 2TiN \text{ (s)} + 8HCl \text{ (g)}, \tag{1}$$

$$2TiCl_4 \text{ (g)} + 4H_2 \text{ (g)} + N_2 \text{ (g)} \rightarrow 2TiN \text{ (s)} + 8HCl \text{ (g)}, \tag{2}$$

where (s) means solid phase and (g) means gas phase. Equation (1) may react at a low temperature of 400–700 °C, but obtain a poor quality of the TiN coating. Conversely, Equation (2) needs a high reaction temperature of 700–1200 °C, but gains a good quality of the TiN coating. Because of the high temperatures of the coking process, and possibly the high pressures, the TiN coating of this work was prepared using the $TiCl_4$–N_2–H_2 system.

The thermal CVD deposition process of TiO_2 coatings was prepared by the $TiCl_4$–H_2–CO_2 system. This is comprised of two homogeneous gas-phase reactions, which can be described by the following chemical reactions:

$$CO_2 \text{ (g)} + H_2 \text{ (g)} \rightarrow CO \text{ (g)} + H_2O \text{ (g)}, \tag{3}$$

$$2H_2O \text{ (g)} + TiCl_4 \text{ (g)} \rightarrow TiO_2 \text{ (s)} + 4HCl \text{ (g)}, \tag{4}$$

where Equation (3) is the homogeneous "water gas shift" reaction to in situ produce the water used in the hydrolysis reaction. Then, $TiCl_4$ reacted with H_2O vapor to gain the TiO_2 coating, as shown in Equation (4).

More details about the preparation of TiN and TiO_2 coatings were described elsewhere [15,16]. It is noteworthy to point out that a preheating furnace was fixed at 650 °C, so that reactants can preheat adequately to the deposition temperature before reaching the surface of substrate, and remove the trace oxygen in system. Moreover, to obtain the growth characteristics and properties of CVD TiN and TiO_2 coatings, the influence of deposition time, residence time, and partial pressure on coating deposition was investigated in detail in the process of experiment. The influence of temperature on TiN and TiO_2 coatings was described in previous works [15,16]. All CVD experiments are based on the following typical experimental condition, as shown in Table 1.

Table 1. A typical experimental condition for the preparation of the TiN and TiO_2 coatings.

Conditions	TiN Coating	TiO$_2$ Coating
Flow rate of H_2 diluent gas (L/min)	1.83	1.88
Flow rate of H_2 carrier gas (L/min)	0.95	0.88
Flow rate of N_2 (L/min)	1.40	none
Flow rate of CO_2 (mL/min)	none	40
Deposition press (atm)	1	1
Preheating temperature (°C)	650	650
Substrate temperature (°C)	850	800
Vaporizer temperature (°C)	28	28
Gas line temperature (°C)	100	100
Deposition time (min)	90	90

The coking test was employed to evaluate the properties of the TiN and TiO_2 coatings. All the tests were carried out in a tubular reactor made of quartz (16 mm i.d. and 350 mm in length), which was placed horizontally inside an electrical furnace. Cyclohexane was used as the cracking feedstock, and the experimental parameters are shown in Table 2. Before the test, the uncoated and TiN-coated specimens were ultrasonically cleaned in acetone and dried in high purity nitrogen (HP–N_2) for 30 min. Then, a specimen was placed in the center of the tube reactor and HP–N_2 passed through the reaction system for 30 min to purge the air inside. Subsequently, the reactor was heated to 770 °C at a rate of 15 °C/min under the HP–N_2 atmosphere protection. The cracking time was fixed at 1.5 h. The reactor was turned off after 1.5 h, and cooled to room temperature in HP–N$_2$. More details about the coking test were described elsewhere [15].

Table 2. Parameters of the coking test.

Conditions	Value
Flow rate of N$_2$ diluent gas (mL/min)	17
Flow rate of N$_2$ carrier gas (mL/min)	73
Substrate temperature (°C)	770
Pre-heater temperature (°C)	400
Vaporizer temperature (°C)	50
Experimental time (h)	1.5
Gas line temperature (°C)	130

After CVD and the coking test, the morphologies of as-deposited and coked TiN and TiO$_2$ coatings were characterized by scanning electron microscopy (SEM) (Hitachi-S-4800, Japan).

3. Results and Discussion

3.1. The Effect of Deposition Time

SEM micrographs of the TiN and TiO$_2$ surfaces with different deposition times are shown in Figure 1. The micrographs indicate that the morphologies of TiN and TiO$_2$ coatings change significantly as the deposition time increases. For the TiN coating, when the deposition time was 5 min, a lot of fine particles intermixed with large spherical particles appeared on the surface of the 310S substrate (Figure 1a). Under higher magnification, it could be seen that these spherical particles consisted of crystallites and showed a rounded agglomerate structure, which was caused by the slow surface mobility of adsorbed reactant molecules and continuous nucleation in this case. However, some original cracks of the 310S substrate were still observed, which suggested that the thickness of the TiN coating after 5 min was not enough to completely cover the surface of the substrate. Similarly, when the deposition time was 30 min, a large number of fine particles, which also consisted of crystallites, appeared on the surface of the substrate (Figure 1b); however, there were fewer surface cracks than on the coating with the deposition time of 5 min. On the one hand, the thickness of the TiN coating increased with the increase of deposition time, and it effectively covered the cracks. On the other hand, with the progress of the deposition process, a secondary nucleation process occurred at the area of the cracks, resulting in the gradual decrease of the cracks. When the deposition time was 90 min, the TiN coating was fully grown to form obvious star-shaped crystals intermixed with lenticular crystals (Figure 1c). The mechanism of formation of star-shaped TiN crystals was proposed by Cheng et al. [27,28] The five extrusion arms of the star-shaped TiN crystals are the result of preferred growth at the five (111) twin boundaries, and the growth rate of these crystals is higher than that of crystals of other shapes.

For the TiO$_2$ coating, when the deposition time was 5 min, it formed irregular particles with a size of approximately 100–1000 nm, and these irregular particles were connected to each other in a plate-like structure (Figure 1d). Meanwhile, the original cracks on the 310S surface disappeared completely with only a few small holes that could be seen under higher magnification. As the deposition time increased from 5 min to 30 min, the small, irregular particles disappeared and were replaced by small hill-like structures with obvious crystal characteristics (Figure 1e). As the deposition time was further increased to 90 min, the morphology of the TiO$_2$ coating changed to a bigger smooth stone-like structure with sharp edges (Figure 1f).

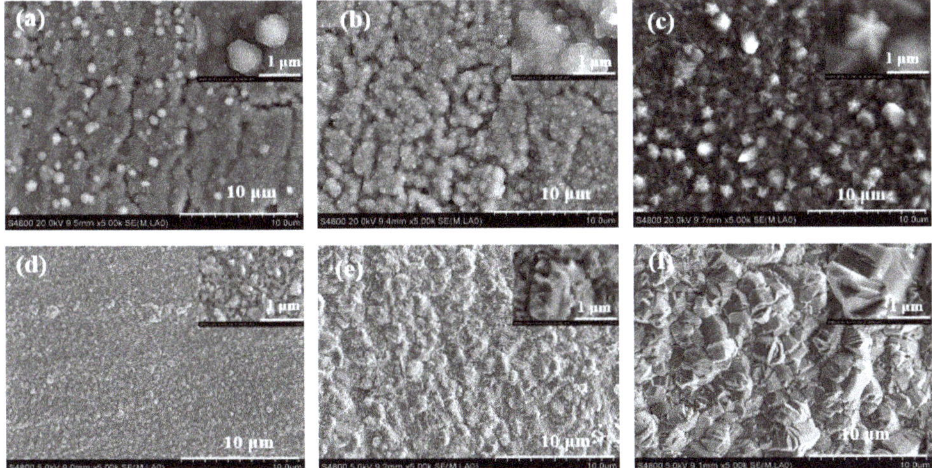

Figure 1. Scanning electron microscopy (SEM) micrographs of TiN at 850 °C with different deposition time: (**a**) 5 min, (**b**) 30 min, and (**c**) 90 min; TiO$_2$ at 800 °C with different deposition time: (**d**) 5 min, (**e**) 30 min, and (**f**) 90 min.

Thus, it can be seen that the effect of deposition time on the growth characteristics of TiN and TiO$_2$ coatings is very different. According to Spiller et al. [29], there are three possible modes of crystal growth on surfaces, namely, the island or Volmer–Weber mode, the layer or Frank–van der Merwe mode, and the layer-plus-island or Stranski–Krastanov mode. In the island (Volmer–Weber) mode, small clusters are nucleated directly on the substrate surface and these grow into islands of the condensed phase. The growth of the TiN coating shows the characteristics of island growth. At the beginning of the deposition, reactant molecules are adsorbed on the surface of the 310S substrate, and then react on the surface to form island nucleation clusters because the atoms (or molecules) of the deposit are more strongly bound to each other than to the substrate. Therefore, many cracks are formed among the islands. As the deposition time increases, because of the growth of the nuclei and the occurrence of secondary nucleation, the islands are linked together and become closely integrated. Subsequently, the cracks on the surface of the TiN coating gradually disappear. In contrast, the growth characteristics of the TiO$_2$ coating can be explained by the layer (Frank–van der Merwe) model, in which the atoms (or molecules) of the deposit are more strongly bound to the substrate than to each other. This mode displays characteristics that contrast with the growth of the TiN coating. In this case, the nuclei of the TiO$_2$ film are sufficiently activated and grow rapidly. The nuclei combine quickly and form a continuous film, which ultimately leads to the disappearance of the cracks on the substrate surface. As the deposition time increases, the TiO$_2$ grain grows gradually, and the crystal structure tends to become complete. However, all three modes of crystal growth are thought to occur on surfaces in the absence of surface defects and interdiffusion and, consequently, the actual situation may be more complicated than the three separate cases.

3.2. The Effect of Residence Time

In this experiment, the residence time was changed by changing the total flow rate without changing the deposition time and partial pressure of all the components. Figure 2 shows the SEM images of TiN coating at 850 °C for 1.5 h at different rates of total flow rate. It can be seen from Figure 2a that when the residence time was long (the total flow rate is 0.14 times as high as that of the typical experiment, as seen in Table 1), the TiN coating also showed cracks, which are typically island growth features, despite the deposition time of up to 90 min. The long residence time means that the reactants can react adequately, nucleate, and grow on the surface of the substrate. However, because of the reduction of the entire reactants in this case, the nuclei did not have sufficient raw materials to grow and form the complete star-shaped TiN crystals. Similar situations occurred at 0.4 times the total flow rate (Figure 2b). The difference is that the particle size of TiN (about 500 nm) in this case was obviously larger than that corresponding to 0.14 times the total flow rate (about 200 nm). When the total flow rate increased to 0.71 times, the TiN coating showed a nodule-like morphology, which was composed of a rounded agglomerate structure and formed by continuous nucleation reaction (Figure 2c). It is worth noting that the star-shaped structure of TiN crystal was visible under higher magnification. As the total flow rate reached 1.28 times, many star-shaped TiN crystals with a size of approximately 1 μm appeared on the substrate surface (Figure 2d).

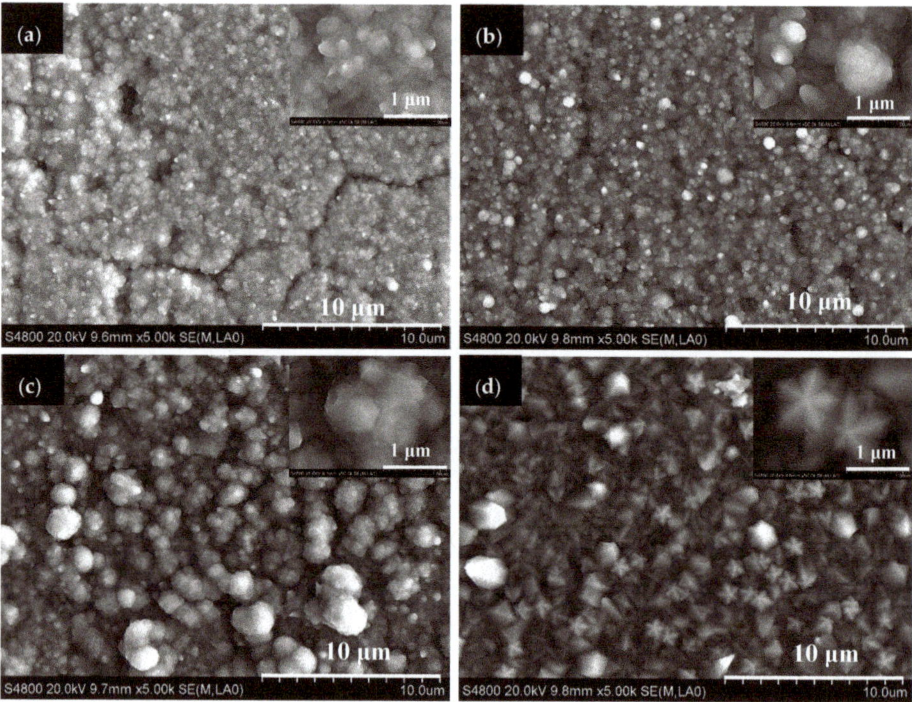

Figure 2. SEM images of TiN coating at 850 °C for 1.5 h at different times of the total flow rate: (**a**) 0.14, (**b**) 0.40, (**c**) 0.71, and (**d**) 1.28 times.

Figure 3 shows the SEM images of the TiO_2 coating at 800 °C for 1.5 h at different multiples of the total flow rate. It is clear from Figure 3 that no cracks appear on the surface of the TiO_2 coating at different multiples of the total flow rate. This result again confirmed that the TiO_2 coating belongs to the layer mode. Interestingly, the morphologies of the TiO_2 coating at 0.4, 0.6, and 1.4 times the total flow rate are very similar, consisting of granular structures with blurred edges (Figure 3a,b,d). It is known that the TiO_2 coating is formed according to Equations (3) and (4). Here, Equation (3) is the homogeneous "water gas shift" reaction, which was investigated comprehensively by Tingey [30]. According to the equation, the water formation rate above 800 °C is given by

$$d[H_2O]/dt = 7.6 \times 10^4 \times e^{-78,000/RT} \times [H_2]^{1/2} \times [CO_2]. \tag{5}$$

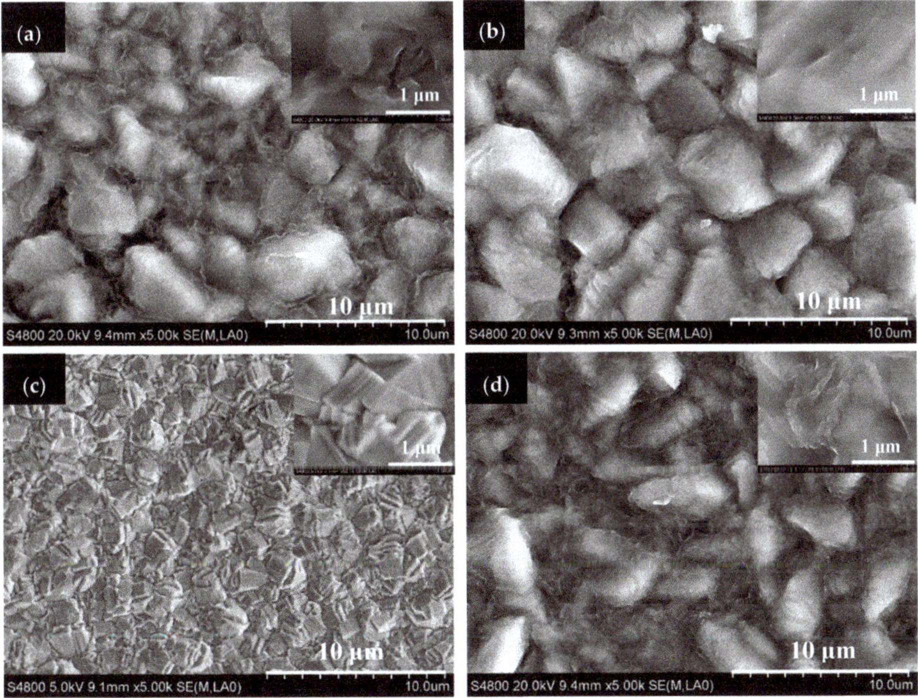

Figure 3. SEM images of the TiO$_2$ coating at 800 °C for 1.5 h at different times of the total flow rate:
(**a**) 0.4, (**b**) 0.6, (**c**) 0.8, and (**d**) 1.4 times.

The reaction represented by Equation (4) is very intense even at room temperature. Therefore, the blurred edges may be the amorphous TiO$_2$, which was formed by rapid reaction and deposition in the gas phase. However, the TiO$_2$ coating formed at 0.8 times the total flow rate (Figure 3c) was very similar to that corresponding to the basic total flow rate (Figure 1f), which shows a smooth stone-like structure with sharp edges. In this case, the TiO$_2$ coating has a better crystal structure. This result indicates that a suitable value of total flow is needed for obtaining a better crystal structure of the TiO$_2$ coating.

3.3. The Effect of Partial Pressure

In this work, the partial pressure was changed by changing the flow rate of each component. Figure 4 shows the SEM images of TiN at 850 °C for 1.5 h at different flow rates of each component. As shown in Figure 4A, as the flow rate of the H$_2$ diluent gas was reduced from 1.83 L/min (base value, Table 1) to 0.95 L/min, the morphologies of the TiN coatings showed nodules (Figure 4(A1)) and the rounded agglomerate structure (Figure 4(A2)). Under a higher magnification, the incompleteness of the star-shaped TiN crystal structure was visible. As shown in Figure 4B, as the flow rate of N$_2$ was reduced from the base value of 1.40 L/min to 0.40 L/min, the star-shaped structure of TiN coatings gradually disappeared, and then presented a disordered sheet structure (Figure 4(B2)). It can be seen that the appropriate flow rate of the H$_2$ diluent gas and N$_2$ is the necessary condition for obtaining the star-shaped structure of the TiN coating. However, as the flow rate of the H$_2$ carrier gas decreased from the base value of 0.95 L/min to 0.20 L/min, the star-shaped structure of the TiN coating first changed to lenticular crystals at 0.65 L/min (Figure 4(C1)), and then changed into larger star-shaped crystals at 0.20 L/min (Figure 4(C2)). Cheng et al. [31] also observed a similar situation in the CVD TiN coating process, and explained that the (111) twin-plane re-entrant corner is the primary growth source in this temperature range for CVD TiN and the extrusion of re-entrant twin boundaries results in plate-like or lenticular crystals. Generally, the microstructure is strongly dependent on the deposition temperature and partial pressure of reaction gas in the CVD process. The influence of the partial pressure may be that the carrier gas contains H_2 and $TiCl_4$, thus it has a double influence on the morphology of the TiN coating.

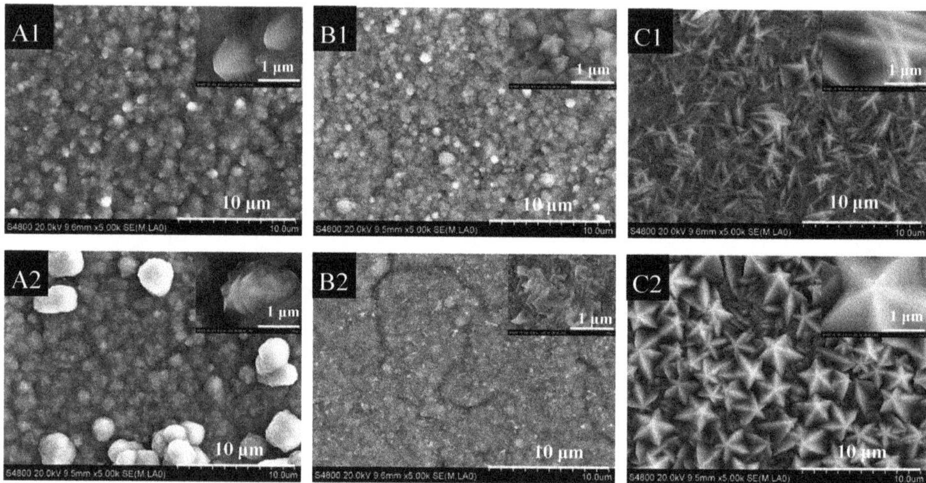

Figure 4. SEM images of TiN at 850 °C for 1.5 h at different flow rate of each component; A group is the flow rate of H_2 diluent gas (L/min): **A1**—1.29, **A2**—0.95; B group is the flow rate of N_2 (L/min): **B1**—1.00, **B2**—0.40; C group is the flow rate of H_2 carrier gas (L/min): **C1**—0.65, **C2**—0.20.

The effects of partial pressures of the H_2 diluent, carrier gas, and CO_2 on the surface morphology of TiO_2 coating at 800 °C are shown in Figure 5. It was clear that the morphology of the TiO_2 coating changed a little with decreasing H_2 partial pressures. This may be because of the high rate of the water gas shift reaction at 800 °C and the severe hydrolysis of $TiCl_4$ (corresponding to Equation (5) and Equation (4)) that makes the TiO_2 coating grow very fast in the layer mode. The effect of decreasing the $TiCl_4$ and CO_2 partial pressures on the morphology of TiO_2 coating is similar to that resulting from the decrease of H_2 partial pressures. The result indicated that the morphology of the TiO_2 coating is a little affected by the partial pressure. In addition, it was found that the size of the TiO_2 particles changed by the CO_2 partial pressure is obviously larger than that obtained by changing the partial pressure of H_2 and $TiCl_4$. This indicates that the growth rate of the TiO_2 grain is most sensitive to the water gas shift reaction.

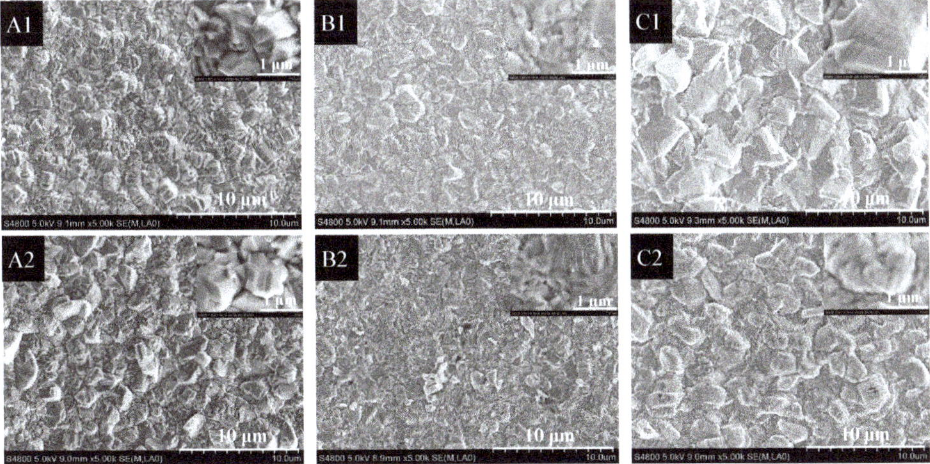

Figure 5. SEM images of TiO$_2$ at 800 °C for 1.5 h at different flow rate of each component; A group is the flow rate of H$_2$ diluent gas (L/min): **A1**—1.52, **A2**—1.32; B group is the flow rate of H$_2$ carrier gas (L/min): **B1**—0.48, **B2**—0.18; C group is the flow rate of CO$_2$ (mL/min): **C1**—80, **C2**—20.

3.4. The Anti-Coking Properties of TiN and TiO$_2$ Coatings

In order to evaluate the effect of growth characteristics on the properties of TiN and TiO$_2$ coatings, coatings with different micromorphologies were selected for the coking test. Figure 6a–d present SEM images (5000:1) of cokes formed on the surface of TiN coatings having different morphologies after cyclohexane steam pyrolysis at 770 °C for 1.5 h. It is obvious that no filamentous cokes formed by metal catalytic coking appear on the surface of all samples. The result demonstrates that the different micromorphology of TiN coatings is uniform and dense, thus it can completely cover the metal substrate and prevent the metal catalytic coking. Meanwhile, carbon deposits are markedly different on the surface of TiN coatings with different morphologies: irregular particles as seen in Figure 6a, needle-like structures as seen in Figure 6b, and fine dust-like particles as seen in Figure 6c,d. When the TiN coating covers the metal substrate surface, the surface reaction of metal catalytic coking is prevented, while the free radical and olefin condensation coking in the bulk phase is dominant. Therefore, reactions in the gas phase are responsible for these structures, and this indicates that the surface microstructure of the coating can affect the structure of carbon deposits.

For the TiO$_2$ coating, the layer-mode growth characteristics specify that the morphology of TiO$_2$ coating is a small change with the change in experimental conditions. To perform the coking tests, samples with significantly different particle sizes were chosen, as shown in Figure 5(A1) and Figure 5(C1). Figure 7a,b present the SEM images (5000:1) of cokes on the surface of TiO$_2$ coatings with different particle sizes that were prepared at different partial pressures. In addition, it was clear that no metal-catalytic filamentous cokes appeared on the TiO$_2$ coating surfaces. Most interestingly, the morphologies of the cokes were very different, as seen in the flaky carbon deposits in Figure 7a and the irregular carbon particles in Figure 7b. A large number of studies have shown that these cokes are amorphous carbon deposits owing to the non-catalytic coking [32,33]. The result suggested that the morphology of non-catalytic cokes is not only affected by the temperature, pressure, and coking precursor, but is also closely related to the surface state of the coatings. In brief, the comparison of the cokes formed on the surface of TiN and TiO$_2$ coatings indicated that both TiN and TiO$_2$ coatings can effectively prevent catalytic coking and eliminate filamentous cokes. Common types of carbon deposits formed by the non-catalytic coking are granular amorphous cokes; however, in some cases, needle-like and flaky carbon deposits are formed, which may be affected by the N or O atoms of the TiN and TiO$_2$ coatings.

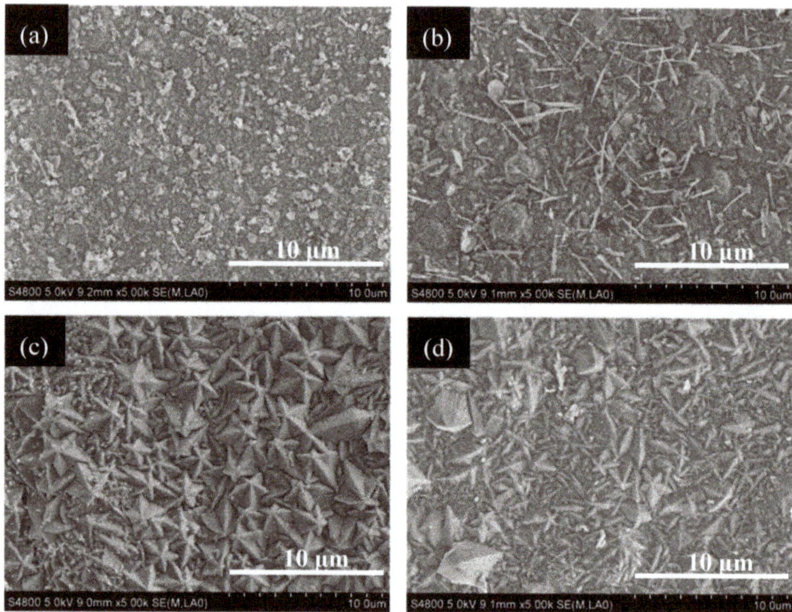

Figure 6. SEM images of carbon deposits on the surface of TiN coatings with different micromorphology after the coking test: (**a**) disordered sheet structure (corresponding to Figure 4(B2)), (**b**) rounded agglomerate structure (corresponding to Figure 4(A2)), (**c**) big star-shaped crystals (corresponding to Figure 4(C2)), and (**d**) lenticular crystals (corresponding to Figure 4(C1)).

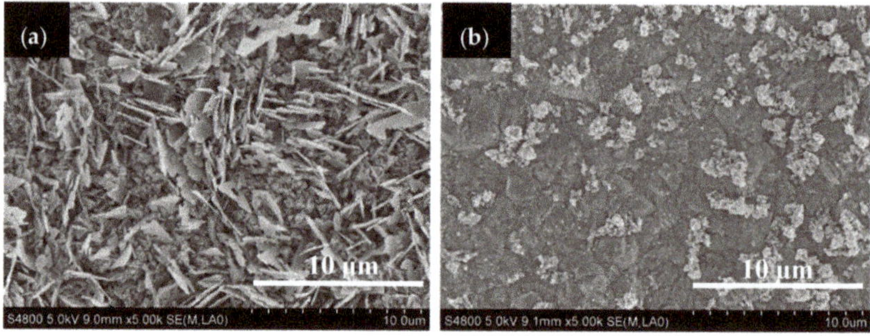

Figure 7. SEM images of carbon deposits on the surface of TiO_2 coatings with different particle sizes after the coking test: (**a**) the flow rate of H_2 diluent gas is 1.52 L/min (corresponding to Figure 5(A1)) and (**b**) the flow rate of CO_2 is 80 mL/min (corresponding to Figure 5(C1)).

4. Conclusions

A series of studies about the growth characteristics and properties of CVD TiN and TiO_2 anti-coking coatings reveal that the effect of deposition parameters on the growth characteristics of TiN and TiO_2 coatings is very different, and it is strongly dependent on deposition time, residence time, and partial pressure. The growth of TiN coating shows the characteristics of the island growth because the atoms (or molecules) of the deposit are more strongly bound to each other than to the substrate. Differently, the growth characteristics of TiO_2 coating growth can be explained by the layer (Frank–van der) mode, which happens when the atoms (or molecules) of the deposit are more strongly bound to substrate than to each other. In general, the growth rate of the star-shaped TiN crystals is higher than that of crystals of other shapes. For the TiO_2 coating, the growth characteristics of the layer mode indicate that the morphology of the TiO_2 coating does not change significantly with the experimental conditions. The size of the TiO_2 particles changed by the CO_2 partial pressure is obviously larger than that obtained by changing the partial pressure of H_2 and $TiCl_4$. This indicates that the growth rate of the TiO_2 grain is most sensitive to the water gas shift reaction.

A coking test was conducted to evaluate the properties of TiN and TiO_2 coatings using cyclohexane steam pyrolysis at 770 °C for 1.5 h. Most interestingly, carbon deposits are very different on the surface of TiN coatings with a different morphology of irregular particles, needle-like structure, and fine dust-like particles. Flaky carbon deposits and irregular carbon particles of cokes on the surface of different particle sizes of TiO_2 coatings that are prepared by different partial pressures are observed. The result suggested that the morphology of non-catalytic cokes is not only affected by temperature, pressure, and coking precursor, but is also closely related to the surface state of the coatings. In summary, from a comparison of the cokes on the surface of the TiN and TiO_2 coatings, both TiN and TiO_2 coatings can effectively prevent catalytic coking and eliminate filamentous cokes. In some cases, however, the N or O atoms in the TiN and TiO_2 coatings may affect common carbon deposits formed by non-catalytic coking, such as formation of needle-like and flaky carbon deposits.

Author Contributions: S.T. and A.T. conceived and designed the experiments; T.L. and J.G. performed the experiments; S.T. and S.D. analyzed the data; A.T. contributed reagents/materials/ analysis tools; S.T. wrote the paper.

Funding: The research was funded by A Plan Project by Department of Science and Technology of Guizhou Province (Qiankehe LH Zi [2016] 7104), Natural science foundation of Guizhou Provincial Department of Education (Qianjiaohe KY Zi [2016] 014), and the Doctoral Scientific Research Foundation of Guizhou Institute of Technology.

Acknowledgments: We would like to sincerely thank Professor Zhu and Professor Wang of Sichuan University for their selfless help. In addition, we should also thank the provincial first-class platform of the Chemical Engineering Practice Teaching Center and the support of College Students' innovation and entrepreneurship.

Conflicts of Interest: The authors declare no conflict of interest.

References

1. Jin, B.; Jing, K.; Liu, J.; Zhang, X.; Liu, G. Pyrolysis and coking of endothermic hydrocarbon fuel in regenerative cooling channel under different pressures. *J. Anal. Appl. Pyrolysis* **2017**, *125*, 117–126. [CrossRef]
2. Edwards, T.; Propul, J. Liquid fuels and propellants for aerospace propulsion: 1903–2003. *Power* **2003**, *19*, 1089–1107. [CrossRef]
3. Cheng, K.; Liang, S.; Huai, X. Effects of chemical heat sink generated by an oxalic acid cooling stream on film-cooling effectiveness. *Exp. Heat Transf.* **2016**, *29*, 113–123. [CrossRef]
4. Wang, H.; Yang, W.; Tian, P.; Zhou, J.; Tang, R.; Wu, S. A highly active and anti-coking Pd-Pt/SiO_2 catalyst for catalytic combustion of toluene at low temperature. *Appl. Catal. A-Gen.* **2017**, *529*, 60–67. [CrossRef]
5. Masirana, N.; Vo, D.V.N.; Salam, M.A.; Abdullah, B. Improvement on coke formation of CaO-Ni/Al_2O_3 catalysts in ethylene production via dehydration of ethanol. *Procedia Eng.* **2016**, *148*, 1289–1294. [CrossRef]
6. Ding, R.; Taylor, M.P.; Chiu, Y.L.; Smith, N.; Mowforth, C.W.; Evans, H.E. Influence of pre-oxidation on filamentary carbon deposition on 20Cr25Ni stainless steel. *Oxid. Met.* **2019**, *91*, 589–607. [CrossRef]
7. Zeng, Z.; Natesan, K. Relationship of carbon crystallization to the metal-dusting mechanism of nickel. *Chem. Mat.* **2003**, *15*, 872–878. [CrossRef]
8. Mohan, R.; Eser, S. Effectiveness of low-pressure metal–organic chemical vapor deposition coatings on metal surfaces for the mitigation of fouling from heated jet fuel. *Ind. Eng. Chem. Res.* **2011**, *50*, 7290–7304. [CrossRef]
9. Yang, C.; Liu, G.; Wang, X.; Jiang, R.; Wang, L.; Zhang, X. Preparation and anti-coking performance of MOCVD alumina coatings for thermal cracking of hydrocarbon fuels under supercritical conditions. *Ind. Eng. Chem. Res.* **2012**, *51*, 1256–1263. [CrossRef]
10. Wang, Z.; Xu, H.; Zhou, J.; Luan, X. Simulation of SiO_2/S coating deposition in a pilot plant set-up for coking inhibition. *Chem. Eng. Res. Des.* **2013**, *91*, 120–133. [CrossRef]

11. Zhou, J.; Wang, Z.; Luan, X.; Xu, H. Anti-coking property of the SiO_2/S coating during light naphtha steam cracking in a pilot plant setup. *Anal. J. Appl. Pyrol.* **2011**, *90*, 7–12. [CrossRef]
12. Zhou, J.; Xu, H.; Liu, J.; Qi, X.; Zhang, L.; Jiang, Z. Study of anti-coking property of SiO_2/S composite coatings deposited by atmospheric pressure chemical vapor deposition. *Mater. Lett.* **2007**, *29*, 5087–5090. [CrossRef]
13. Bao, B.; Liu, J.; Xu, H.; Liu, B.; Zhang, W. Inhibitory effect of $MnCr_2O_4$ spinel coating on coke formation during light naphtha thermal cracking. *RSC Adv.* **2016**, *6*, 68934–68941. [CrossRef]
14. Tang, S.; Gao, S.; Hu, S.; Wang, J.; Zhu, Q.; Chen, Y.; Li, X. Inhibition effect of APCVD titanium nitride coating on coke growth during n-hexane thermal cracking under supercritical conditions. *Ind. Eng. Chem. Res.* **2014**, *53*, 5432–5442. [CrossRef]
15. Tang, S.; Gao, S.; Wang, S.; Wang, J.; Zhu, Q.; Chen, Y.; Li, X. Characterization of CVD TiN coating at different deposition temperatures and its application in hydrocarbon pyrolysis. *Surf. Coat. Technol.* **2014**, *258*, 1060–1067. [CrossRef]
16. Tang, S.; Wang, J.; Zhu, Q.; Chen, Y.; Li, X. Preparation of rutile TiO_2 coating by thermal chemical vapor deposition for anti-coking applications. *ACS Appl. Mater. Interfaces* **2014**, *6*, 17157–17165. [CrossRef]
17. Tang, S.; Shi, N.; Wang, J.; Tang, A. Comparison of the anti-coking performance of CVD TiN, TiO_2 and TiC coatings for hydrocarbon fuel pyrolysis. *Ceram. Int.* **2017**, *43*, 3818–3823. [CrossRef]
18. Altin, O.; Eser, S. Analysis of solid deposits from thermal stressing of a JP-8 fuel on different tube surfaces in a flow reactor. *Ind. Eng. Chem. Res.* **2001**, *40*, 596–603. [CrossRef]
19. Baker, R.T.K.; Chludzinski, J.J.; Lund, C.R.F. Further studies of the formation of filamentous carbon from the interaction of supported iron particles with acetylene. *Carbon* **1987**, *25*, 295–303. [CrossRef]
20. Reyniers, M.S.G.; Froment, G.F. Influence of metal surface and sulfur addition on coke deposition in the thermal cracking of hydrocarbons. *Ind. Eng. Chem. Res.* **1995**, *34*, 773–785. [CrossRef]
21. Crynes, L.L.; Crynes, B.L. Coke formation on polished and unpolished Incoloy 800 coupons during pyrolysis of light hydrocarbons. *Ind. Eng. Chem. Res.* **1987**, *26*, 2139–2144. [CrossRef]
22. Marek, J.C.; Albright, L.F. Formation and removal of coke deposited on stainless steel and vycor surfaces from acetylene and ethylene. *ACS Symp.* **1982**, *202*, 123–149.
23. Marek, J.C.; Albright, L.F. Surface phenomena during pyrolysis: The effects of treatments with various inorganic gases. *ACS Symp.* **1982**, *202*, 151–175.
24. Gregg, S.J.; Leach, H.F. Reaction of nickel with carbon monoxide at elevated temperatures. *J. Catal.* **1966**, *6*, 308–313. [CrossRef]
25. Durbin, M.J.; Castle, J.E. Carbon deposition from acetone during oxidation of iron—the effects of chromium and nickel. *Carbon* **1976**, *14*, 27–33. [CrossRef]
26. Tang, S.; Wang, J.; Zhu, Q.; Chen, Y.; Li, X. Oxidation behavior of CVD star-shaped TiN coating in ambient air. *Ceram. Int.* **2015**, *41*, 9549–9554. [CrossRef]
27. Cheng, H.; Lin, T.; Hon, M. Multiple twins induced preferred growth in TiN and SiC films prepared by CVD. *Scripta Mater.* **1996**, *35*, 113–116. [CrossRef]
28. Cheng, H.; Hon, M. Growth mechanism of star-shaped TiN crystals. *J. Cryst. Growth* **1994**, *142*, 117–123. [CrossRef]
29. Venables, J.A.; Spiller, G.; Hanbucken, M. Nucleation and growth of thin films. *Rep. Prog. Phys.* **1992**, *47*, 5–29.
30. Tingey, G.L. Kinetics of the water-gas equilibrium reaction. I. the reaction of carbon dioxide with hydrogen. *J. Phys. Chem.* **1966**, *70*, 1406–1412. [CrossRef]
31. Cheng, H.E.; Chiang, M.J.; Hon, M.H. Growth characteristics and properties of TiN coating by chemical vapor deposition. *J. Electrochem. Soc.* **1995**, *142*, 1573–1578. [CrossRef]
32. Xie, W.; Fang, W.; Li, D.; Xing, Y.; Guo, Y.; Lin, R. Coking of model hydrocarbon fuels under supercritical condition. *Energy Fuels* **2009**, *23*, 2997–3001. [CrossRef]
33. Eser, S.; Venkataraman, A.R.; Altin, O. Deposition of carbonaceous solids on different substrates from thermal stressing of JP-8 and jet-A fuels. *Ind. Eng. Chem. Res.* **2006**, *45*, 8946–8955. [CrossRef]

Article

Comparative Study of the Performances of Al(OH)$_3$ and BaSO$_4$ in Ultrafine Powder Coatings

Weihong Li [1], Diego Cárdenas Franco [2], Marshall Shuai Yang [2], Xinping Zhu [3], Haiping Zhang [1], Yuanyuan Shao [1], Hui Zhang [1,2,*] and Jingxu Zhu [1,2]

[1] Collaborative Innovation Center of Chemical Science and Engineering (Tianjin), School of Chemical Engineering and Technology, Tianjin University, Tianjin 300072, China; weihli@tju.edu.cn (W.L.); hpzhang@tju.edu.cn (H.Z.); yshao@tju.edu.cn (Y.S.); jzhu@uwo.ca (J.Z.)

[2] Particle Technology Research Center, Department of Chemical & Biochemical Engineering, The University of Western Ontario, London, ON N6A 5B9, Canada; cardenas_franco@yahoo.ca (D.C.F.); marshall.yang@uwo.ca (M.S.Y.)

[3] Wesdon-Rivers Institute of Powder Coatings Science and Technology, Zhaoqing 526000, China; xzhu269@gmail.com

* Correspondence: hzhang1@tju.edu.cn; Tel.: +86-222-349-7607

Received: 7 May 2019; Accepted: 23 May 2019; Published: 27 May 2019

Abstract: Ultrafine powder coatings are one of the development directions in the powder coating industry, as they can achieve thin coatings with good leveling and high surface smoothness comparable to liquid coatings. Compared to regular coatings, they experience a higher sensitivity to any incompatibilities, e.g., filler from coating components. The properties of fillers play a great role in the performance of coating films. Aluminum trihydrate (Al(OH)$_3$) is a well-known filler in solvent-based coatings and other polymer industries. To study and evaluate the performances of Al(OH)$_3$ in ultrafine powder coatings, a popular filler, barium sulfate (BaSO$_4$) is used for comparison. Both fillers are added in ultrafine powder coatings based on two of the most commonly used resin systems (polyester-epoxy and polyester). The differences of physical and chemical properties between both fillers have significant influences on several properties of powder paints and coating films. The polar groups (hydrogen bond) in Al(OH)$_3$ result in the strong interaction between inorganic filler and organic polymer matrix, thus decreasing the molecular network mobility and influencing the chain formation, which is verified by differential scanning calorimetric (DSC). The bed expansion ratio (BERs) of powder paints incorporated with Al(OH)$_3$ are much higher than those with BaSO$_4$, which indicate more uniform gas-solid contact during the spraying process. Samples with Al(OH)$_3$ exhibit much lower specular gloss at 60°, which are expected to achieve remarkable matting effects. Superior corrosion resistances can be observed for almost all the coated panels incorporated with Al(OH)$_3$ in contrast to those with BaSO$_4$. Other aspects are slightly influenced by the difference between the two fillers, such as the angle of repose values (AORs) of powder paints, the impact resistance and flexibility of coating films.

Keywords: Al(OH)$_3$; BaSO$_4$; filler; ultrafine powder coatings

---

1. Introduction

During the last few decades, engineers and researchers have attached great importance to powder coatings due to their economic and environmental benefits. Compared with solvent-borne coatings, they eliminate the use of volatile organic compounds (VOC) that are both expensive and environmentally unfriendly. Moreover, the overspray paint powders can be reclaimed and reused, resulting in a nearly 100% transfer efficiency [1,2]. However, the particle sizes of regular powder coating are generally in the range from 30μm to 60μm, which leads to a thicker film with rougher appearance in comparison with solvent-borne coatings.

Recently, there has been a strong trend to reduce the particle size to improve the surface quality of the final coatings and lead to the considerable cost savings of material and energy [2–7]. Therefore, ultrafine powder coatings (D_{50} < 25 μm) have attracted more and more attention from the finishing industries, such as oral drug delivery systems, dental and orthopedic implants and the pharmaceutical industry [8–10]. In contrast to regular powder coatings (D_{50} > 30 μm), the ultrafine powder coatings are with a higher sensitivity to any incompatibilities from coating components and application environment due to their smaller mean particle size, poorer flowability and the thinner film they can form [1,2,7]. Considering their potential utilization and sensitivity, study on the effects of coating components, e.g. fillers on their performance, is more valuable compared to the regular powder coatings. By including fillers in the ultrafine powder coatings, the agglomerates formed may cause a decrease in its flowability as well as defects during spraying and curing process [2,7,11]. Therefore, the study of the effects of filler addition on the performance of the ultrafine powder is of great significance.

Fillers are one of five principal components of powder coatings that include polymer resins, curing agents (also called hardeners or cross-linkers), pigments and additives. They are usually inorganic substances with chemical stability and are produced artificially or naturally like minerals. Fillers have two functions in coatings: one is to reduce costs by decreasing the dosage of the polymer resin, normally the main-cost part in the formulation; the other is to modify certain physical, chemical or visual properties of the coating [12–16]. The commonly used fillers in powder coatings include blanc fixe ($BaSO_4$), lithopone ($ZnS·BaSO_4$), talc ($Mg_3Si_4O_{10}(OH)_2$), zinc white (ZnO) and carbon black (C), etc. [17–22]. Above all, $BaSO_4$ is used in a high percentage of powder paints because of its electric conductivity and light transparency in thin coatings.

$Al(OH)_3$ is an extensively used filler in liquid coatings, and its usage in the polymer industry has been reported [23–26]. Owing to its higher thermal conductivity, it can be used in room temperature vulcanized (RTV) silicone rubber coatings for the improvement of tracking and corrosion resistances [27]. Several researchers and $Al(OH)_3$ suppliers have also noted the superiority of $Al(OH)_3$ in improving the mechanical and other properties of powder coatings [28–30]. However, little systematic research is so far reported on the comparisons of the effects of $Al(OH)_3$ and $BaSO_4$ on the physical properties, ultraviolet (UV) and corrosion resistances in powder coatings.

In this study, $Al(OH)_3$ and $BaSO_4$ were compared as fillers in the most widely used polyester-epoxy (hybrid) and polyester ultrafine powder coatings. The paper investigated the effects of $Al(OH)_3$ as well as $BaSO_4$ on flowability of powder coatings, mechanical and other physical properties of coating films, plus their performances under corrosive environments and other external circumstances. In addition, the results were discussed regarding the difference of physical and chemical properties between both fillers.

2. Experimental

2.1. Preparation and Characterization of Powder Paints

As described earlier, powder coatings consist of five principal components, including polymer resins, curing agents, pigments, additives and fillers. The materials used in this study are shown in Table 1. In this research, four different filler contents were tested in two widely used resin systems. Each sample contains different type and content of resins and curing agents, as shown in Table 2. To determine the effects of type and content of fillers, the other components (pigments, degassing agents and flow agents) remained constant. Physical properties of fillers incorporated in the coatings are shown in Table 3.

Table 1. Materials.

Components	Hybrid (Polyester-Epoxy)	Polyester
Resin	Crylcoat 2440-2	Crylcoat 2440-2
Curing agent	DER 663U	TGIC/Araldite PT 810
Filler	$Al(OH)_3$/Custom Grinders Polyfill 301 $BaSO_4$/Sparwite W-10	
Pigment	Carbon black/Raven 5000 Ultra II	
Degassing agent	Benzoin/S602	
Flow agent	Acrylic polymer/Lanco P10	
Fluidization additive	SiO_2/AEROSIL R972	

Hybrid (polyester: epoxy = 7:3 wt.%).

Table 2. Formulations of powder paints.

Specimen	Resin Type	Filler	Contents/(wt.%)		
			Resin + Curing Agent	Filler	Others
H-C		-	97.6	0	
H-A1/H-B1	Hybrid	$Al(OH)_3$/$BaSO_4$	80.3	17.3	
H-A2/H-B2			70	30.6	
H-A3/H-B3			57.4	40.2	2.4
PE-C		-	97.6	0	
PE-A1/PE-B1	Polyester	$Al(OH)_3$/$BaSO_4$	80.3	17.3	
PE-A2/PE-B2			70	30.6	
PE-A3/PE-B3			57.4	40.2	

Others: pigment 1.4 wt.%, degassing agent 0.3 wt.%, flow agent 0.7 wt.%.

Table 3. Physical properties of fillers used in the study.

Filler	Shape	Median Particle Size D_{50}/(μm)	Specific Gravity/(-)	Color	Refractive Index/(-)	Oil Absorption/ (g oil/100 g)
$BaSO_4$	Aggregates	2.1	4.4	White	1.64	10
$Al(OH)_3$	Irregular	9	2.4	White	1.58	28

As illustrated in Figure 1, the appropriate materials were premixed and then extruded by a twin-screw extruder (SLJ-10, Donghui Powder Coating Processing Equipment Co., Ltd., Yantai, China). The hot extrudates were cooled, crushed into small chips and added into an air classifier mill (ACM 02, Donghui Powder Coating Processing Equipment Co., Ltd., Yantai, China) for fine grinding, classifying and screening. The median particle size (D_{50}) of paint powders was within the range of 18 μm ± 0.5 μm, as shown in Table 4, which was measured by a laser particle size analyzer (BT2000B, Bettersize instruments Ltd., Dandong, China). Microstructure and morphology of the powder paints and coating films were characterized with scanning electron microscopy (SEM, S4800, Hitachi High-Technologies Global, Tokyo, Japan). Differential scanning calorimetric (DSC) was used to determine the glass transition temperature (T_g) of coatings and $\Delta H_{crosslinking}$ of the crosslinking events. The DSC curves were obtained by a calorimeter (DSC-3, Mettler Toledo Inc., Greifensee, Switzerland) at temperatures between 25 °C and 210 °C. It was operated at a heating rate of 10 °C/min under inert nitrogen atmosphere (N_2) with a flow rate of 50 mL/min.

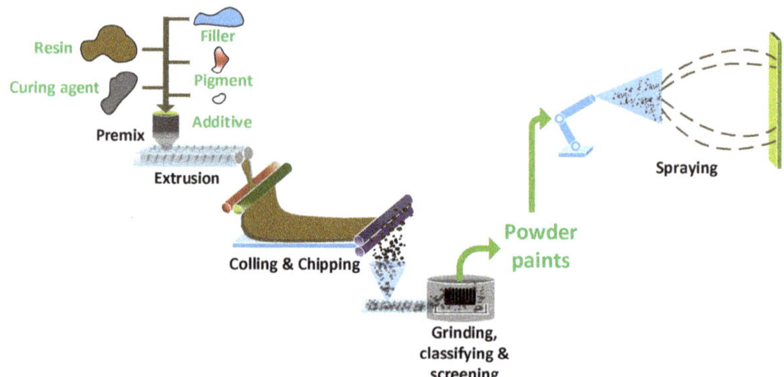

Figure 1. Schematic diagram of the manufacturing process of powder coating.

Table 4. Particle size of powder paints.

Specimen	Particle Size/(μm)			Specimen	Particle Size/(μm)		
	D_{10}	D_{50}	D_{90}		D_{10}	D_{50}	D_{90}
H-C	5.77	18.02	60.02	H-C	5.77	18.02	60.02
H-A1	5.29	18.29	60.71	H-B1	5.49	18.11	59.64
H-A2	5.12	18.40	60.14	H-B2	5.71	18.18	60.6
H-A3	5.48	18.10	58.33	H-B3	5.68	17.79	59.01
PE-C	5.29	17.82	59.64	PE-C	5.29	17.82	59.64
PE-A1	5.30	18.54	59.99	PE-B1	5.51	18.24	59.70
PE-A2	5.91	18.04	58.64	PE-B2	5.79	18.13	59.17
PE-A3	5.83	17.71	56.79	PE-B3	5.66	18.10	58.96

According to Geldart's Powder Classification, the ultrafine powder paints used in this study can be categorized in Group C (particle size under 25 μm–30 μm), which is difficult to fluidize due to its cohesive nature [31]. The cohesion is attributed to the relatively larger interparticle attractive forces, which can be overcome by the addition of nanoparticles (fluidization additives), thus improving the flowability of Group C particles [32–35]. Fluidization additive used in this study is listed in Table 1, which is dry blended (0.7 wt.%) before the powder paints are applied onto the substrates.

Flow behavior of powder paints was characterized with angle of repose (AOR) and bed expansion ratio (BER) tests [36]. As shown in Figure 2, AOR is defined as the angle between the horizontal and the side of the cone when the powder falls freely on the plate. It reflects the cohesiveness and internal friction of a powder and is commonly employed to characterize the flow behavior of powder samples. In general, powder samples with AOR over 45° are usually considered cohesive, while samples with a smaller AOR have better flowability [37].

BER is a useful parameter that indicates the extent to which powders expand at a specific superficial gas velocity. It is the ratio of the bed height at operating conditions (expanded) to the fixed bed height, as defined in Equation (1). Generally, BER is used to characterize fluidization quality based on the belief that a higher BER indicates a better gas-solid contact [38]. A higher BER implies a large amount of gas trapped among the paint powders, which contributes to adequate gas-solid contact during the following spraying process, resulting in a smoother visual appearance. In this study, fluidization of the paint powders was conducted in a homemade fluidized bed as schematically shown in Figure 3. The fluidized bed column was made of Plexiglas that was 75 cm tall with I.D. of 5 cm. The expanded bed height was measured simultaneously when the paint powders were fluidized steadily at each set of superficial gas velocity. In this study, the BER is reported using a gas velocity ranging from 0 cm/s to near 1.0 cm/s, which is representative fluidization in a powder coating process [1,39].

$$BER = \frac{H_{expanded\ bed}}{H_{fixed\ bed}} \tag{1}$$

The specific gravity is a significant parameter for calculating the cost benefit of powder coatings, usually in the form of *square meters/kg powder coating* in a given film thickness. It can provide theoretical usage and coverage of powder coatings.

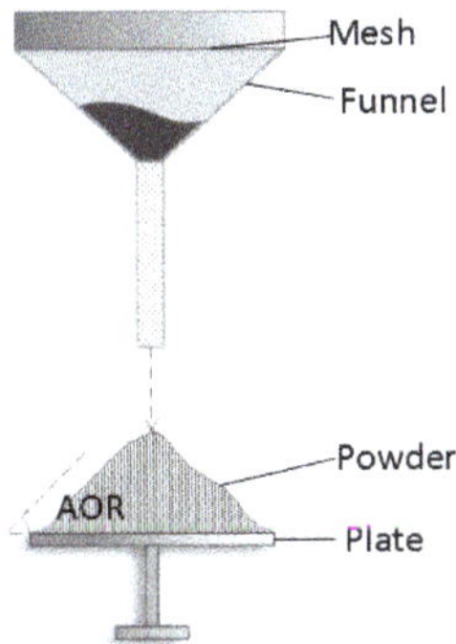

Figure 2. Schematic diagram of AOR measurement.

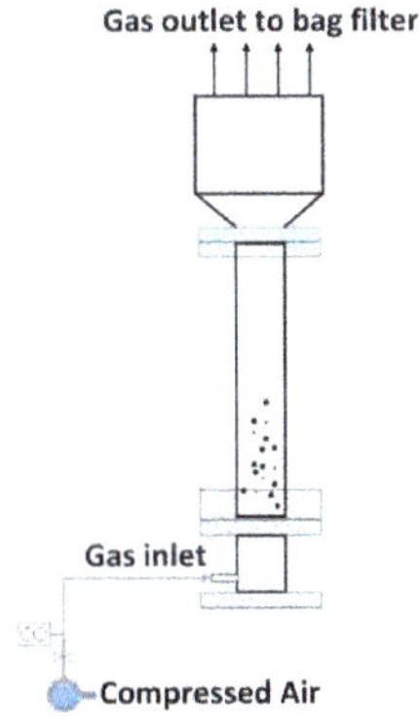

Figure 3. Schematic diagram of fluidized bed.

In this study, all the powder paints were fabricated by an ACM. The operating parameters are virtually the same, with minor adjustments to ensure all the paint samples' D_{50} fall in the range of 18 µm ± 0.5 µm and are in the same particle shape. Consequently, the differences of AORs and BERs caused by particle size and shape can be neglected. The characterizations mentioned above were performed by devices and instruments listed in Table 5.

Table 5. Measurements of powder paints and coating films.

Measurements	Apparatuses	ASTM Standards
Powder paints		
AOR	PT-X (Hosokowa Micron Corporation, Hirakata, Japan)	-
Specific gravity	Volumetric flask Balance (XSE105DU, Mettler Toledo Inc., Greifensee, Switzerland)	D5965-02 (2013)
BER	Homemade fluidized bed	-

Table 5. *Cont.*

Measurements	Apparatuses	ASTM Standards
Coating films		
Pencil scratch hardness	PH5800 (BYK Additives & Instruments Ltd., Wesel, Germany)	D3363-05 (2011)
Surface roughness	SJ-210 (Mitutoyo Co., Kanagawa, Japan)	D4417-2014
Impact resistance	QCJ 120 (HANYI instruments Co. Ltd., Wuhan, China)	D2794-93 (2010)
Flexibility	BYK 5750 (BYK Additives & Instruments Ltd., Wesel, Germany)	D522M-13
Specular gloss	IQ206085 (Rhopoint Components Ltd., East Grinstead, UK)	D523-14 (2018)
Salt spray test	MX-9204 (Associated Environmental Systems Ltd., Hong Kong)	B117-2011, D1654-08 (2016)
UV accelerated test	XE-3 (Q-Lab Corporation Ltd., Cleveland, OH, USA)	G155-2013, D523-14 (2018)

2.2. Preparation and Characterization of Finished Films

After measurements of powder paints described before, each of the powder samples was sprayed on two different types of panels from Q-Lab (Q-Lab Corporation Ltd., Westlake, OH, USA): aluminum panel (88.9/63.5/0.8, L/W/T, mm) and steel panel (127.0/76.2/0.8, L/W/T, mm). All panels were degreased with acetone prior to spraying. Powder samples were sprayed using a Sure coat corona gun (Nordson Corporation, Westlake, OH, USA). All the sprayed panels were cured at 204 °C for 10 min to ensure complete curing. After curing, a film thickness of 38 μm ± 5% were retained for all the subsequent measurements. The thickness of the coating film was measured by a PosiTector 6000 thickness gage (DeFelsko Corporation, Ogdensburg, NY, USA).

The performance of coating films depends on many factors; for instance, the properties of the resin systems, environment exposed and other factors. Many test methods for developing and monitoring film properties are available to simulate the application conditions and accelerate the degradation process of powder coatings. In this study, for evaluating the pencil scratch hardness, impact resistance, conical mandrel bend test, specular gloss, UV and corrosion resistances of the film surface by instrumental measurements, detailed evaluation procedures are followed according to appropriate ASTM standards. Characterizations of the coating films were conducted 24 h after cured. Instruments used and ASTM standards followed in the measurements are presented in Table 5.

Pencil scratch hardness test uses constant pressure and special pencils (from 9H to 9B degrees) to scratch the finished film to determine the hardness of the film. The Impact resistances of the coated films were characterized by an impact tester. A steel weight with a hemispherical head (diameter: 13 mm, weight: 1.8 kg) was dropped from a height between 20 and 120 cm onto the coated panels. The same procedure is repeated by increasing the height by 25 mm increments until cracks appear, which presents the failure of the coating. Flexibility of the coating films is measured by a conical mandrel bend apparatus. The panels are bent for about 135°. The radius of the cone varies from 1.5 mm to 19 mm. The coating film is deemed flexible if no crack appears. Specular gloss describes the power of a test surface to reflect light specularly. A complete specular light reflection is 100, while a complete diffuse reflection is 0. The surface roughness test is conducted to characterize the discontinuities over some distance (15 mm in this study) on the specimen surface. To ensure the reliability and reproducibility of the test results, the measurements were performed in three different regions for each of the finished films. The salt spray test is an essential method to evaluate corrosion resistance of a coating film in terms of loss of adhesion under long-term exposure to salt fog. An X-form-scribe was made on the surface of the coated panel leaving the metal substrate exposed. The representative mean rust creepages can be determined by a rating grade from 10 to 1 (Table 6). The resistance of a powder coated film against UV exposure (UV accelerate test) is one of the most important parameters to determine the applicability of a particular coating for external conditions. The UV tester used in this study is equipped with two xenon-arc lamps with a maximum wavelength of 340 nm. A cycle consists of 102 min of UV irradiation (0.35 W/m^2, 63 °C) in dry conditions and 18 min in water spray conditions. The measured specular gloss at 60° was used to calculate the gloss retention of the coating film. The gloss retention is defined as:

$$\text{Gloss retention} = \frac{\text{Gloss}_{t=x}}{\text{Gloss}_{t=0}} \tag{2}$$

Table 6. Rating number of failures at scribe.

Mean Rust Creepage/(mm)	Rating Number
Zero	10
0 to 0.5	9
0.5 to 1.0	8
1.0 to 2.0	7
2.0 to 3.0	6
3.0 to 5.0	5
5.0 to 7.0	4
7.0 to 10.0	3
10.0 to 13.0	2
13.0 to 16.0	1
>16.0	0

3. Results

3.1. Flow Behavior and Specific Gravity of Paint Powders

Flow behaviors of the powder paints were characterized by AOR and BER measurements. Figure 4 shows the AORs of the powder samples dry-blended with fluidization additives with respect to the fillers' types and contents. It is clear that the type of resin system affects the AORs of the paint samples. AORs of the samples based on polyester are much higher than that of the hybrid. However, for the same resin system, the type of filler only makes a small difference in AOR, if the same amount of Al(OH)$_3$ and BaSO$_4$ are used. As mentioned before, the particle size distribution, shape and humidity of powder samples can be ignored because of the same manufacturing processes, operating parameters and conditions, as shown in Figure 5. The slightly higher AORs of samples with BaSO$_4$ is partially due to its smaller D$_{50}$, as shown in Table 3, which results in a larger tendency to the formation of agglomerates and poor dispersion in powder paints.

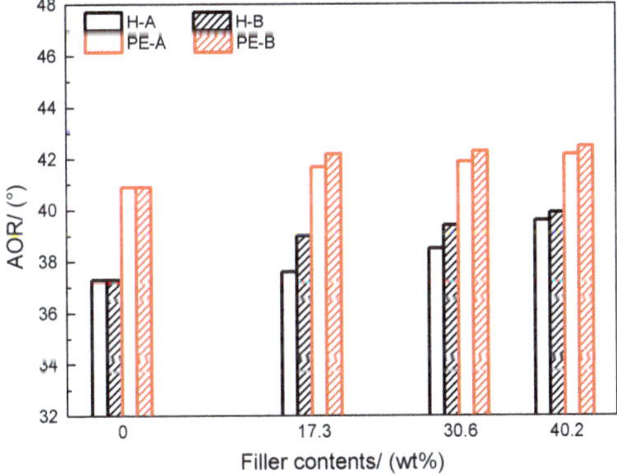

Figure 4. AORs of samples.

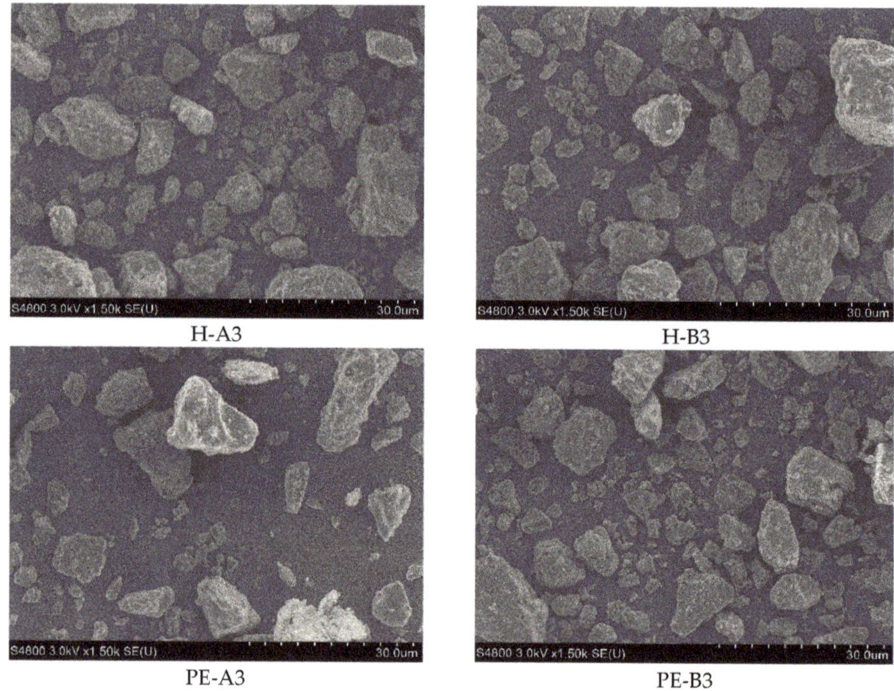

H-A3   H-B3

PE-A3   PE-B3

Figure 5. SEM micrographs of powder paints.

From our prior paper and industrial experience, when AORs are higher than 42°, the paint powders tend to agglomerate and thus exhibit poor flowability during the spraying process [40]. Accordingly, the filler contents in this study for $Al(OH)_3$ and $BaSO_4$ are quite acceptable for hybrid-based coatings with the addition of fluidization additives; the AORs of most of the paint samples vary below 42°, except for those based on polyester resins.

BER was used to characterize flow behavior based on the belief that a higher BER indicates more gas in the interstitial void among particles, implying more uniform gas-solid contact and thus better flow behavior and fluidization quality. Figure 6 presents the result of BER measured from samples based on hybrid and polyester. Evidently, filler-free powder paints exhibit the highest BERs in almost the whole range of superficial gas velocity. With the increase of filler contents, BERs of powder paints deteriorate and decrease to about 1.8 and 1.7 for samples of H-B3 and PE-B3, respectively. From previous investigations, the BERs for regular powder coatings (D_{50} >30μm) showed a maximum value of 1.6 when the superficial gas velocity was 1.0 cm/s [1]. Compared to regular powder coatings, the ultrafines normally exhibit poorer flowabilities, such as lower BER and higher AOR, because of the stronger inter-particle forces [32–36], while with the aid of the fluidization additives the BERs of the ultrafine powder coatings in this research vary from 1.7 to 2.4 at a similar superficial gas velocity of 0.93 cm/s (Figure 7). A higher BER leads to a better gas-solid contact, which results in a better film appearance and gas saving in the electrostatic spraying process.

Besides, PE-powders are much sensitive to filler's incorporation, BERs of polyester powders decrease more significantly in contrast to those of hybrid. With the increase of filler contents, the BERs of PE-B and PE-A deteriorate very fast and decrease by more than 30% and 20% in contrast to that of PE-C, while the reduction of the BERs of H-B and H-A are 19% and 10%, respectively.

Figure 8 shows the specific gravities of paint powders respect to different contents of two fillers. It is observed that all samples possess similar specific gravities of 1.2 when no filler is incorporated.

Obviously, the addition of fillers into the powder coatings directly raise its specific gravity and the samples with the same filler contents have close values. The overall density of samples incorporated with $BaSO_4$ is much higher than that of $Al(OH)_3$, which is due to the difference of specific densities between the two fillers.

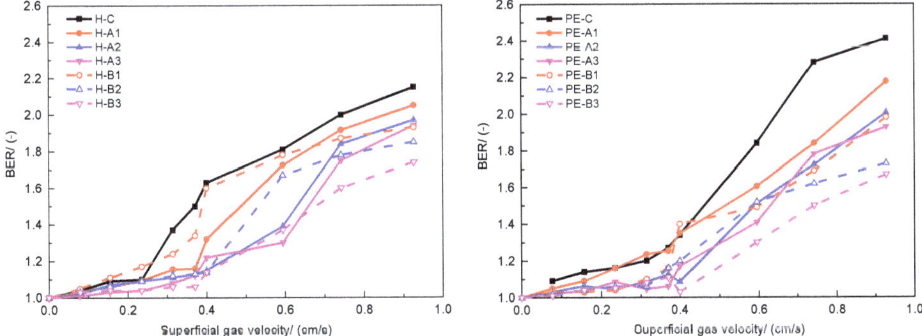

Figure 6. Bed expansion ratios (BERs) of samples respect to superficial gas velocity.

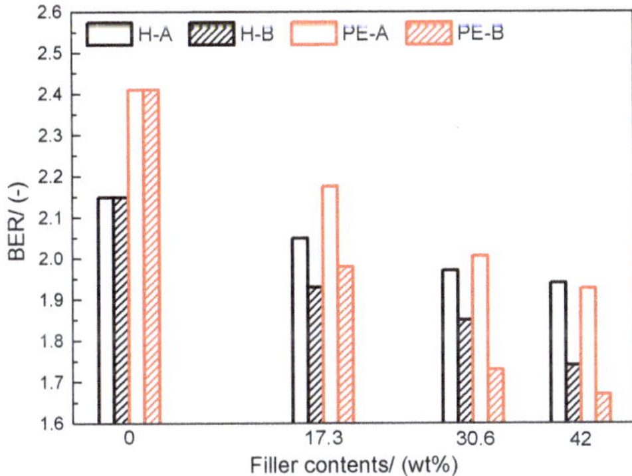

Figure 7. BERs of samples at gas velocity of 0.93 cm/s.

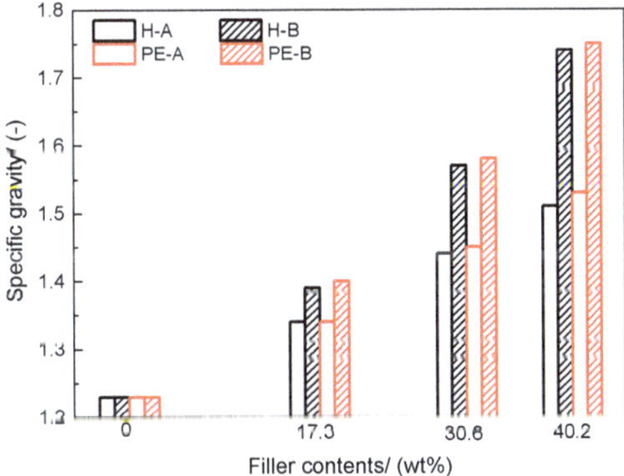

Figure 8. Specific gravities of samples.

3.2. Physical Properties, UV and Corrosion Resistances of Coating Films

3.2.1. Physical Properties

Surface hardness, impact resistance, and flexibility are all crucial properties for typical protective coating films. Figure 9 exhibits the pencil scratch hardness test comparisons of samples respect to fillers' types and contents. As expected, owing to the similar resin and curing agent, samples exhibit similar pencil scratch hardness of HB when no filler is incorporated in. It is observed that the pencil scratch hardness of samples with $Al(OH)_3$ increase to H when the filler content is 30.6%. While, for samples incorporated with $BaSO_4$, the pencil scratch hardness increased to H at the maximum $BaSO_4$ loading. Compared to $BaSO_4$, $Al(OH)_3$ is more efficient to increase the hardness of coating film at higher loadings. The H-A3 sample exhibited the highest hardness of 2H.

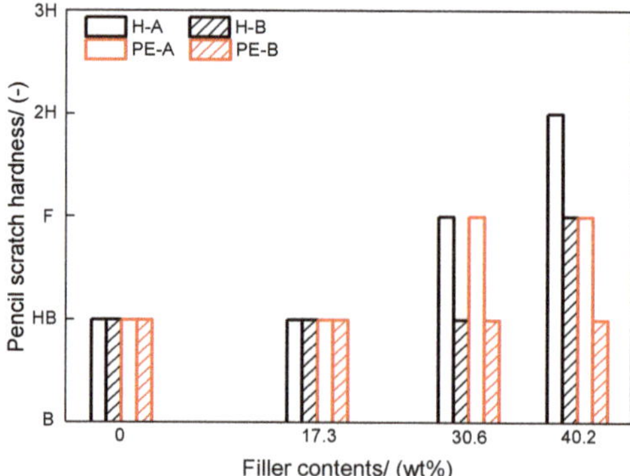

Figure 9. Pencil scratch hardness of samples.

Figure 10 exhibits the results of the impact resistance for samples in respect to filler types and contents. It is observed that the addition of fillers caused a minor decrease in the impact resistance of the coating films. This may be due to the discontinuities in the film matrix introduced by inorganic fillers, which makes the film less flexible and therefore lose adhesion to the panels when receiving an impact. $Al(OH)_3$ shows a slight influence on the impact resistance of coatings. For samples incorporated with $Al(OH)_3$, the decrease of impact resistance occurs at the maximum filler loading (40.2%), with the reduction of impact resistance for about 1.1 J (from 21.2 J to 20.1 J). As for samples added with $BaSO_4$, the reduction of impact resistance starts at a filler content of 17.3%.

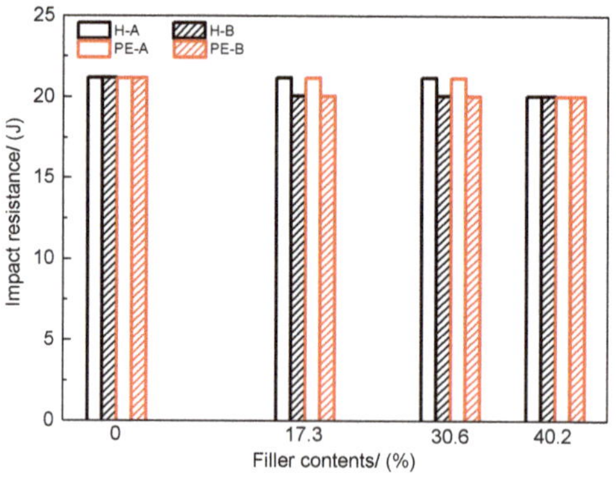

Figure 10. Impact resistances of samples.

The coating flexibility was tested with the conical mandrel bend device. The results show that samples exhibit good flexibility up to filler content of 40.2%, and there is no sign of cracking when the bending exceeds 3 mm mandrel diameter, which is the minimum radius of the cone used in this study.

In general, specular gloss is one of the most commonly used parameters for evaluating surface optical quality. Figure 11 presents the result of the evaluation of specular gloss at 60°. When no filler is incorporated, the specular gloss varies from 95.8 (PE-C) to 98.4 (H-C). The increase of filler contents leads to reduction of the specular gloss. At the maximum load of fillers, the specular glosses decrease by about 14.3% (H-B3) and 21.9% (PE-B3), while they fall further down from 47.8% (H-A3) and 49.5% (PE-A3) in contrast to H-C and PE-C. At the same filler content, samples containing $Al(OH)_3$ exhibit much lower levels of gloss than those with $BaSO_4$. The specular glosses of H-A and PE-A decrease significantly to about 50 at the maximum loadings of $Al(OH)_3$, while samples H-B and PE-B show glossy films with a specular gloss of about 80.

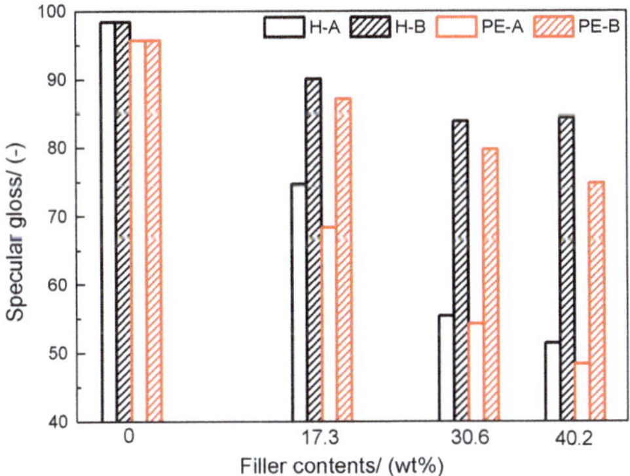

Figure 11. Specular gloss of samples with different fillers and filler contents at 60°.

3.2.2. UV and Corrosion Resistances

Ultraviolet light degrades the coatings by imparting energy into the films. The energy of UV light destroys the crosslink by generating heat or breaking chemical bonds. The heat and breakage of bonds will lead to the loss of physical and chemical properties of coating films, bringing in chalking, color change and gloss reduction.

Figure 12 presents the gloss retention at 60° after a UV accelerated test against 1000 h of UV exposure. PE based samples exhibit better overall UV resistances in contrast to H ones when no filler is incorporated in. The gloss retention of hybrid based samples decreased rapidly to 90% of the initial gloss within 200 h. Polyester-based samples have the best UV resistances, the gloss retention keeps 90% of the initial gloss for about 700 h in the test chamber, which is 3.5 times of that of H based samples.

Besides, it is observed that with the use of $Al(OH)_3$, the PE-based films exhibit much higher gloss retentions than those with $BaSO_4$, as shown in Figure 13. Compared with $Al(OH)_3$, the gloss retentions of most samples incorporated with $BaSO_4$ deteriorate significantly with the increase of filler contents. Samples PE-A1 and PE-B1 keep 90% of the initial gloss for about 630 h and 530 h against UV exposure. Sample PE-A2 keeps 90% of initial gloss for more than 600 h, which is 1.5 times of that of PE-B2. When increasing the fillers' loading to the maximum, the gloss retention of PE-A3 decreases quickly, which is partially due to the much more significant consumption of resin and curing agents in samples incorporated with $Al(OH)_3$. A similar trend can be observed in samples based on hybrid resin systems.

A salt spray test provides a method by which to evaluate the corrosion resistance in terms of loss of adhesion at a scribe mark. The rust creepages for scribed samples can be evaluated by rating grade between 10 and 1 (Table 6). Figure 14 presents the results of corrosion resistance of samples with different fillers and filler contents. Samples filled with $Al(OH)_3$ exhibit better overall corrosion resistance especially for H-A samples, which keeps rating number of grade 9 throughout the whole tested filler contents, while the rating number of H-B samples decrease to grade 7 when the filler contents exceed 30.6 wt.%. The corrosion resistances of samples added with $BaSO_4$ deteriorate significantly with the increase of fillers contents, especially at the maximum loadings, except for H-A samples.

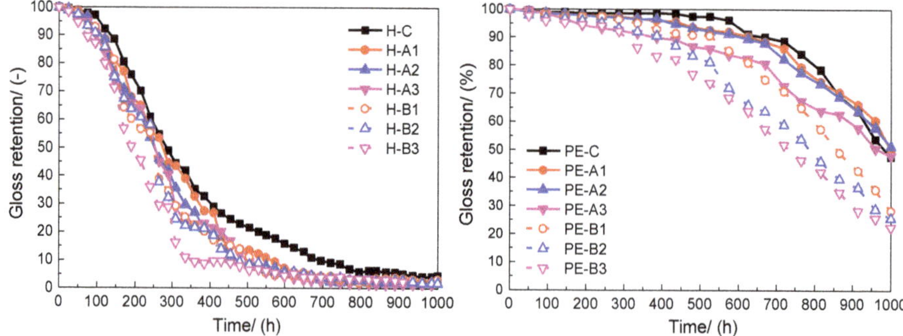

Figure 12. Gloss retention of samples after 1000 h of UV accelerate test.

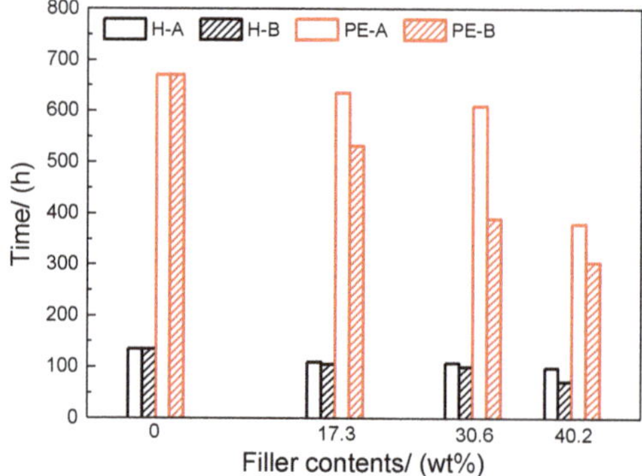

Figure 13. Time consumed of samples at 90% of gloss retention.

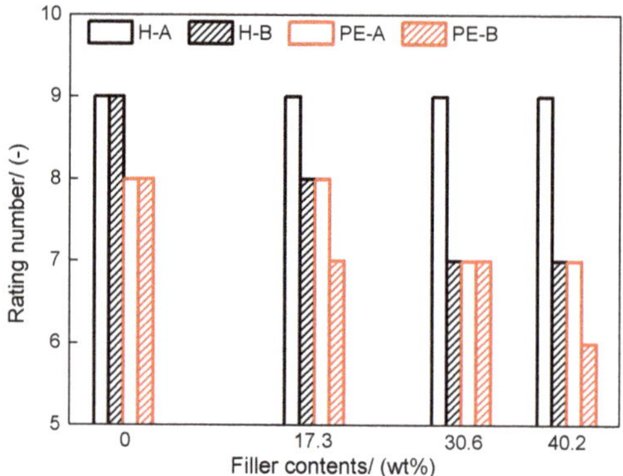

Figure 14. Rating number of failures at scribe of samples after 1000 h of salt spray test.

4. Discussion

This study comparatively investigated the effect of $Al(OH)_3$ and $BaSO_4$ on performances of ultrafine powder coatings. The results show that the incorporation of $Al(OH)_3$ or $BaSO_4$ make little difference to several properties of the powder paints and finished films, such as AORs of powders, the flexibility and impact resistance of coating films, while the other properties are influenced significantly by the fillers' types and contents. It is shown that the BERs of powders filled with $Al(OH)_3$ is much higher than those of $BaSO_4$. The specular gloss of coatings incorporated with $BaSO_4$ is much higher in contrast to those with $Al(OH)_3$. Coatings with $Al(OH)_3$ exhibit outstanding corrosion resistance performances when exposed to salt fog, especially for hybrid-based coatings even at higher filler content.

4.1. The Effect of Fillers' Physical Properties on Performances of Powder Paints and Coating Films

The relatively smaller median particle size, higher specific gravity of $BaSO_4$ and higher oil absorption of $Al(OH)_3$ may partially result in the difference of BER and specular gloss between coating films filled with $Al(OH)_3$ and $BaSO_4$ respectively.

Samples incorporated with $Al(OH)_3$ exhibit higher BERs in contrast to those containing $BaSO_4$ at 0.93 cm/s of air velocity, as shown in Figure 7. As described earlier, $BaSO_4$ has a much higher specific gravity than $Al(OH)_3$. Generally, the higher the specific gravity of particles in the fluidized bed, the lower bed expansion is achieved at the same air velocity.

The specular glosses of finished films decreased significantly with the addition of $Al(OH)_3$ in contrast to those of $BaSO_4$, indicating a stronger matting effect from $Al(OH)_3$. A matte surface can scatter the light in all directions and hence the surface appears less glossy. A micro-roughness surface is necessary for creating the diffuse light scattering responsible for the visual effects of reduced gloss. Compared with $BaSO_4$, the better matting effect of $Al(OH)_3$ incorporated coatings are partially due to the higher oil absorption of $Al(OH)_3$ particles (28g oil/100 g), which is almost 3 times of that of $BaSO_4$ (10g oil/100 g). During the curing reaction, the non-uniform shrinkage in micro scale which results from the cross-linking of binder materials, together with the presence of many filler particles occurs. In general, the higher the oil absorption value of the filler, the more binder it requires to bind it. Compared with $BaSO_4$, the naturally higher oil absorption of $Al(OH)_3$ particles lead to the increase of the content of $Al(OH)_3$ particles in the top-surface of the films and thus the formation of a micro-rough surface during the curing process. The micro-roughness produced is identified by SEM images and profile of surface roughness. Figure 15 shows the cross-section view of coating films PE-A3 and PE-B3. The polymer matrix appears darker than the aluminum substrate, as indicated by the white line which marks out the interface of substrate and the coating film. The coating film adheres tightly to the substrate, without obvious defects or pores at the interface. The thickness of the coating film is approximately 38 μm. Figure 16 exhibits the surface roughness of coating films PE-A3 and PE-B3. The average roughness Ra is the arithmetic average value of the surface profile throughout the length of the testing surface. The Ra of PE-A3 is more than 1.5 times of that of PE-B3, which is due to the micro non-uniformity produced during the curing process.

In general, powder coatings are matted by the help of matting agents. They are incompatible with the coating and can produce micro-roughness of the surface scattering the incident light in different directions. Generally, matting agents are either polyolefinic waxes or inorganic fillers that can decrease the gloss level to about 40% of the initial ones when used in concentrations about 4%. However excessive use of wax can produce haze, yellowing or oily appearance [11,12]. The addition of $Al(OH)_3$ can enhance the matting effect without the incorporation of any other matting agents. The specular gloss is achieved to blow 50 with the sample (PE-A3) at the maximum loading of $Al(OH)_3$. In this case, $Al(OH)_3$ is a good alternative to achieve lower gloss finishes in one shot without the necessity of adding matting agents.

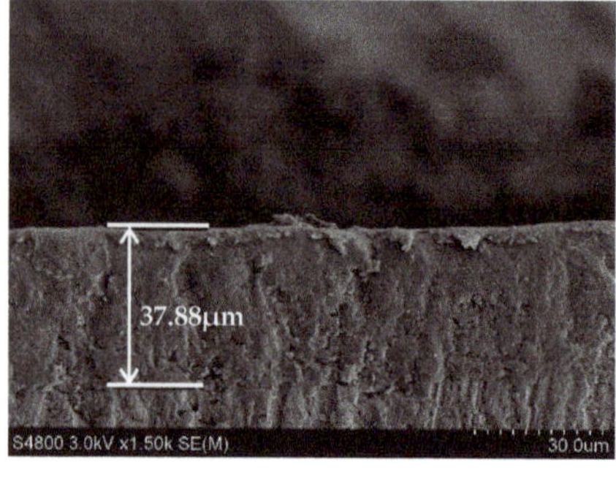

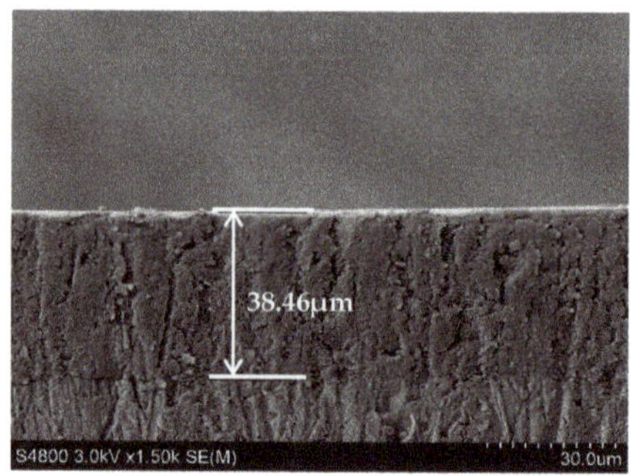

(**a**) PE-A3-cross-section view   (**b**) PE-B3-cross-section view

Figure 15. SEM micrographs of coating films with the addition of $Al(OH)_3$ and $BaSO_4$ respectively.

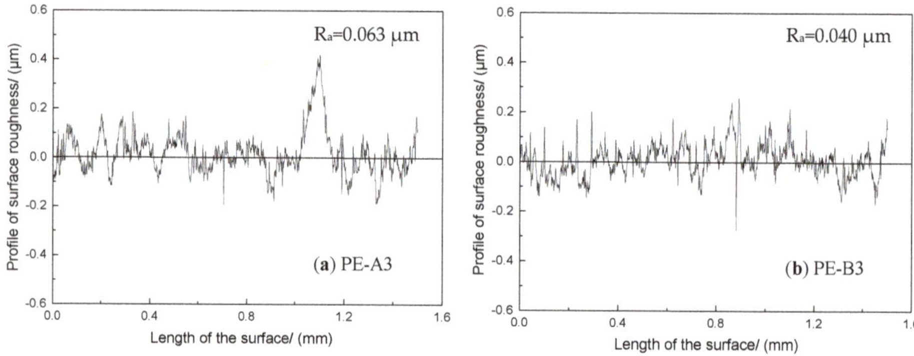

(**a**) PE-A3   (**b**) PE-B3

Figure 16. Roughness of the surface of coating films with the addition of $Al(OH)_3$ and $BaSO_4$ respectively.

4.2. The Effect of Fillers' Chemical Properties on Performances of Coating Films

The chemical properties of fillers make for a great influence on the properties of coating films, due to the interaction between polymer and fillers during the curing processes. The differential scanning calorimetry (DSC) analysis can reveal the effect of chemical properties on the coatings.

Figure 17 shows the DSC analysis of coatings H-C, H-B3 and H-A3. Two important events of the curing processes of coatings, the glass transition and the crosslinking are identified. At first, the glass transition curves developed much slower at early stages of curing process when the molecular chains formed. Then the glass transition accelerated at later stages in accordance with the chain network formed [41]. With the incorporation of $BaSO_4$ and $Al(OH)_3$, the T_g increases from 62.1 °C (H-C) to 63.9 °C (H-B3) and 64.6 °C (H-A3). This is partially due to the interaction of the polymer resin systems and the inorganic fillers, which hinder the molecular mobility during the curing processes [42]. Besides, the amount of the crosslinking heat released ($\Delta H_{crosslinking}$) during the scan were greatly decreased by the addition of two fillers in this study. During the curing process, the fillers absorbed parts of the energy depending on the filler quantity in the resin systems, which is also confirmed by other investigations [42–44]. The dispersion of inorganic fillers in the resin systems hinders the mass and heat transfer, decreasing the reactivity of the system as evidenced by the reduction in the $\Delta H_{crosslinking}$ of the crosslinking processes, as shown in Figure 17. The fillers' addition significantly decreased the amount of heat released. The $\Delta H_{crosslinking}$ of H-A3 and H-B3 are about 39% and 37% of the heat released by sample H-C, which are much lower than the polymer content of 59.8% in H-A3 and H-B3. Meanwhile, fillers with polar groups such as hydrogen bond can interact with the polymeric network, thus influencing the chain formation and flexibility during the curing process, consequently increasing the T_g, decreasing the $\Delta H_{\text{crosslinking}}$. The H-A3 coatings filled with $Al(OH)_3$ exhibit higher T_g and lower $\Delta H_{\text{crosslinking}}$ than those of H-B3, which is partially due to the -OH bond interacting with the polymer chain.

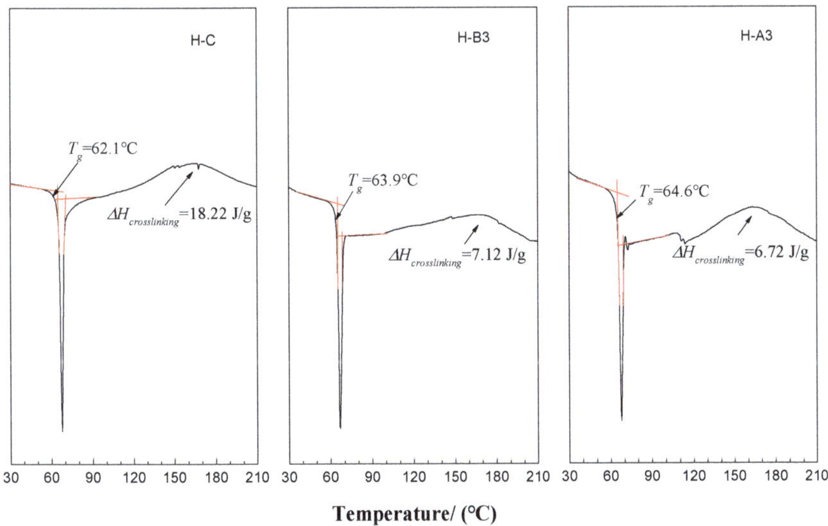

Figure 17. DSC thermograms of hybrid-based coatings with and without the addition of fillers.

The glass transition temperature (T_g) is a useful parameter to describe curing propagation and can be linked to mechanical properties [45]. During the curing processes, the molecular network mobility decreases according with the increase of T_g [44]. This can partially explain the significantly improved pencil scratch hardness of H-A3 and H-B3.

Besides, the corrosion resistance is also greatly influenced by the chemical properties of fillers added. It is a crucial property for the coating's application, which strongly depends on a variety of parameters such as quality of resin systems, chemical properties of bulk materials and conditions the films exposed to. The reason that leads to the superior corrosion resistances of H-A could be interpreted as follows. Firstly, the existence of epoxy in hybrid components, a well-known corrosion resistance resin, helps maintaining a good anti-corrosion property. Secondly, a large amount of OH bonds are exposed to the surface of the $Al(OH)_3$ particles which can interact with the resin chains through hydrogen bond [46], as shown in Scheme 1. As a result, higher corrosion resistances of samples H-A1, H-A2, H-A3 and PE-A1 are obtained in contrast to those added with $BaSO_4$. With the increase of filler contents, the deterioration of corrosion resistance is mainly due to the significantly inadequate of resin and curing agent, which are essential in crosslink inside coatings [47,48].

Scheme 1. Hydrogen bonding between hybrid resin and $Al(OH)_3$.

5. Conclusions

$Al(OH)_3$, a widely used filler in solvent-based coatings and $BaSO_4$, a popular filler in both solvent-based coatings and powder coatings, are involved in polyester-epoxy and polyester based ultrafine powder coatings to make comparative investigations.

1. Several performances of the products are significantly influenced by the type of fillers, such as BERs and specific gravity of powder paints and specular gloss, pencil scratch hardness and corrosion resistance of coating films. Powder paints with $Al(OH)_3$ exhibit higher BERs at high superficial gas velocity than those with $BaSO_4$, ensuring a uniform gas-solid contact during the electrostatic spraying process. A significant promotion in matting effect is shown in samples with $Al(OH)_3$. It shows great enhancement in pencil scratch hardness of samples added with $Al(OH)_3$ in contrast to those with $BaSO_4$. PE based samples incorporated with $Al(OH)_3$ show excellent UV resistances. Hybrid based samples with $Al(OH)_3$ demonstrate outstanding performance of corrosion resistance.

2. Other properties of powder paints and coating films are slightly influenced by the difference of fillers' type, such as the AORs, the flexibility and the impact resistance. Compared to the filler's type, the resin system has greater influence on AORs of powder paints. With the addition of fluidization additives, the AORs of most hybrid-based samples stay within an acceptable range ($<42°$), and the powders can be expected to have good flowability during the application process. Flexibility of films remain stable for sample panels with both fillers. Impact resistance of samples incorporated with $Al(OH)_3$ is slightly superior than those with $BaSO_4$.

3. The different performances of coating films with the two fillers are attributed to the difference of their physical and chemical properties. The more significant matting effect of $Al(OH)_3$ compared to $BaSO_4$, is mainly due to the relatively higher oil absorption and micro-rough surface produced accordingly. The excellent corrosion resistances of hybrid-based samples incorporated with $Al(OH)_3$ are partially due to that the hydrogen bond of $Al(OH)_3$ strengthens the crosslink of coating films. The DSC analysis reveals relatively higher T_g and slightly lower $\Delta H_{crosslinking}$ of H-A3 samples, which are partially due to the strong interaction between the hydrogen bond and the polymer chain.

This experimental study provides valuable information for the application of $Al(OH)_3$ to ultrafine powder coatings. Compared to $BaSO_4$, the use of $Al(OH)_3$ is expected to bring about a better mechanical property and durability of the coating film.

Author Contributions: Conceptualization, J.Z. and H.Z.; Methodology, D.C.F.; Software, M.S.Y.; Validation, H.Z., H.Z. and W.L.; Formal analysis, W.L.; Investigation, W.L. and D.C.F.; Resources, H.Z. and Y.S.; Data curation, W.L.; Writing—original draft preparation, W.L. and X.Z.; Writing—review and editing, H.Z. and H.Z.; Visualization, W.L.; Supervision, H.Z.; Project administration, J.Z.; Funding acquisition, H.Z.

Funding: This research was funded by the Natural Sciences and Engineering Research Council of Canada (NSERC), Discovery Grant RGPIN-2018-06256.

Conflicts of Interest: The authors declare no conflicts of interest.

Notations

D_{50}	Median particle diameter (50 vol.% of the powder smaller than the diameter) (μm)
AOR	Angle of repose, which is defined as the slope measured from the horizontal to the side of a cone that is formed when the powder is allowed to free fall onto a plate. It is the maximum angle at which powders can pile up (°)
BER	Bed expansion ratio, which is the ratio of the bed height at operating conditions to the finally settled bed height
$Gloss_{t=0}$	Initial gloss at 60° (-)
$Gloss_{t=x}$	Gloss value at a general time "x" at 60° (-)

References

1. Jesse, Z.; Hui, Z. Ultrafine powder coatings: An innovation. *Powder Coat.* **2005**, *16*, 39–47.
2. Misev, T.A.; Van der Linde, R. Powder coatings technology: New developments at the turn of the century. *Prog. Org. Coat.* **1998**, *34*, 160–168. [CrossRef]
3. Monsalve-Bravo, G.M.; Bhatia, S.K. Modeling Permeation through Mixed-Matrix Membranes: A Review. *Processes* **2018**, *6*, 172. [CrossRef]
4. Satoh, H.; Harada, Y.; Libke, S. Spherical particles for automotive powder coatings. *Prog. Org. Coat.* **1998**, *34*, 193–199. [CrossRef]
5. Biris, A.S.; Mazumder, M.K.; Yurteri, C.U.; Sims, R.A.; Snodgrass, J.; De, S. Gloss and Texture Control of Powder Coated Films. *Part. Sci. Technol.* **2001**, *19*, 199–217. [CrossRef]
6. Meschievitz, T.; Rahangdale, Y.; Pearson, R.U.S. council for automotive research (USCAR) low-emission paint consortium: A unique approach to powder painting technology development. *Met. Finish.* **1995**, *93*, 26–31. [CrossRef]
7. Zhang, H.; Yang, M.S.; Bhuiyan, M.T.I.; Zhu, J. Green Chemistry for Automotive Coatings: Sustainable Applications. In *Green Chemistry for Surface Coatings, Inks and Adhesives: Sustainable Applications*; Royal Society of Chemistry, Thomas Graham House: Cambridge, UK, 2019; Chapter 3; ISBN 978-1-78262-994-8.
8. Yang, Q.; Ma, Y.; Shi, K.; Yang, G.; Zhu, J. Electrostatic coated controlled porosity osmotic pump with ultrafine powders. *Powder Technol.* **2018**, *333*, 71–77. [CrossRef]
9. Hou, N.Y.; Zhu, J.; Zhang, H.; Perinpanayagam, H. Epoxy resin-based ultrafine dry powder coatings for implants. *J. Appl. Polymer Sci.* **2016**, *133*. [CrossRef]
10. Yang, Q.; Ma, Y.; Zhu, J. Sustained drug release from electrostatic powder coated tablets with ultrafine ethylcellulose powders. *Adv. Powder Technol.* **2016**, *27*, 2145–2152. [CrossRef]
11. David, H. Powder Coatings. In *The Technology, Formulation and Application of Powder Coatings*; Sussex, J.D.S.W., Ed.; John Wiley and Sons Ltd.: Chichester, West Sussex, England, 2000; Volume 1, ISBN 978-0471978992.
12. Liberto, N. *User's Guide to Powder Coating*, 4th ed.; Association for Finishing Processes and Society of Manufacturing Engineers: Dearborn, MI, USA, 2003.
13. Yang, G.; Heo, Y.-J.; Park, S.-J. Effect of Morphology of Calcium Carbonate on Toughness Behavior and Thermal Stability of Epoxy-Based Composites. *Processes* **2019**, *7*, 178. [CrossRef]
14. Cummings, S.; Zhang, Y.; Smeets, N.; Cunningham, M.; Dubé, M.A. On the Use of Starch in Emulsion Polymerizations. *Processes* **2019**, *7*, 140. [CrossRef]
15. George, W. *Handbook of Fillers*, 4th ed.; ChemTec Publishing: Toronto, ON, Canada, 2016.
16. Dongmei, Z.; Fa, L.; Liangming, X.; Wancheng, Z. Preparation and properties of glass coats containing SiCN nano powder as filler. *Mater. Sci. Eng. A* **2006**, *431*, 311–314.
17. Luo, S.; Zheng, Y.; Li, J.; Ke, W. Effect of curing degree and fillers on slurry erosion behavior of fusion-bonded epoxy powder coatings. *Wear* **2003**, *254*, 292–297. [CrossRef]
18. Karthik, S.; Siva, P.; Balu, K.S.; Suriyaprabha, R.; Rajendran, V.; Maaza, M. Acalypha indica–mediated green synthesis of ZnO nanostructures under differential thermal treatment: Effect on textile coating, hydrophobicity, UV resistance, and antibacterial activity. *Adv. Powder Technol.* **2017**, *28*, 3184–3194. [CrossRef]
19. Chen, C.; Qiu, S.; Cui, M.; Qin, S.; Yan, G.; Zhao, H.; Wang, L.; Xue, Q. Achieving high performance corrosion and wear resistant epoxy coatings via incorporation of noncovalent functionalized graphene. *Carbon* **2017**, *114*, 356–366. [CrossRef]
20. Qingyan, S.; Sue, H.; Weili, W.; Dongsheng, F.; Tianlong, M. Preparation and characterization of antistatic coatings with modified BaTiO$_3$ powders as conductive fillers. *J. Adhes. Sci. Technol.* **2013**, *27*, 2642–2652.
21. Feng, Q.; Ming, W.; Yongbo, H.; Shaoyun, G. The effect of talc orientation and transcrystallization on mechanical properties and thermal stability of the polypropylene/talc composites. *Compos. Part A* **2014**, *58*, 7–15.
22. Shafiq, M.; Yasin, T.; Aftab Rafiq, M. Structural, thermal and antibacterial properties of chitosan/ZnO composites. *Polym. Compos.* **2014**, *35*, 79–85. [CrossRef]
23. Hamdani-Devarennes, S.; Longuet, C.; Sonnier, R.; Ganachaud, F.; Lopez-Cuesta, J.M. Calcium and aluminum-based fillers as flame-retardant additives in silicone matrices. III. Investigations on fire reaction. *Polym. Degrad. Stab.* **2013**, *98*, 2021–2032. [CrossRef]

24. Xanthos, M. *Functional Fillers for Plastics*; Wiley-VCH: Weinheim, Germany, 2005.
25. Cardenas, M.A.; Garcia-Lopez, D.; Gobernado-Mitre, I.; Merino, J.C.; Pastor, J.M.; Martinez, J.D.; Barbeta, J.; Calveras, D. Mechanical and fire retardant properties of EVA/clay/ATH nanocomposites—Effect of particle size and surface treatment of ATH filler. *Polym. Degrad. Stab.* **2008**, *93*, 2032–2037. [CrossRef]
26. Nasir, K.M.; Sulong, N.H.R.; Johan, M.R.; Afifi, A.M. An investigation into waterborne intumescent coating with different fillers for steel application. *Pigment Resin Technol.* **2018**, *47*, 142–153. [CrossRef]
27. Seyedmehdi, S.A.; Zhang, H.; Zhu, J. Superhydrophobic RTV silicone rubber insulator coatings. *Appl. Surf. Sci.* **2014**, *258*, 2972–2976. [CrossRef]
28. Li, W.; Franco, D.C.; Yang, M.S.; Zhu, X.; Zhang, H.; Shao, Y.; Zhang, H.; Zhu, J. Investigation of the Performance of ATH Powders in Organic Powder Coatings. *Coatings* **2019**, *9*, 110. [CrossRef]
29. Huber Engineered Materials. Available online: https://www.hubermaterials.com/products/flame-retardants-smoke-suppressants/flame-retardant-smoke-suppressant-applications/industrial-and-paper-coatings/powder-coatings.aspx (accessed on 17 February 2019).
30. Alumina Chemicals & Castables. Available online: http://www.aluminachemicals.net/alumina-powder-4288120.html (accessed on 17 February 2019).
31. Geldart, D. Types of gas fluidization. *Powder Technol.* **1973**, *7*, 285–292. [CrossRef]
32. Turki, D.; Fatah, N.; Saidani, M. The impact of consolidation and interparticle forces on cohesive cement powder. *Int. J. Mater. Res.* **2015**, *106*, 1258–1263. [CrossRef]
33. Bruni, G.; Lettieri, P.; Newton, D.; Barletta, D. An investigation of the effect of the interparticle forces on the fluidization behaviour of fine powders linked with rheological studies. *Chem. Eng. Sci.* **2007**, *62*, 387–396. [CrossRef]
34. Visser, J. Van der Waals and other cohesive forces affecting powder fluidization. *Powder Technol.* **1989**, *58*, 1–10. [CrossRef]
35. Chen, Y.; Yang, J.; Dave, R.N.; Pfeffer, R. Granulation of cohesive Geldart group C powders in a Mini-Glatt fluidized bed by pre-coating with nanoparticles. *Powder Technol.* **2009**, *191*, 206–217. [CrossRef]
36. Lim, K.S.; Zhu, J.X.; Grace, J.R. Hydrodynamics of gas-solid fluidization. *Int. J. Multiph. Flow* **1995**, *21*, 141–193. [CrossRef]
37. Cheremisinoff, N.P.; Cheremisinoff, P.N. *Hydrodynamics of Gas-Solids Fluidization*; Gulf Publishing Co.: Houston, TX, USA, 1984.
38. Huang, Q. Flow and fluidization properties of fine powders. Ph.D. Thesis, Philosophy-University of Weston Ontario, London, ON, Canada, 2009.
39. Krantz, M.; Zhang, H.; Zhu, J. Characterization of powder flow: Static and dynamic testing. *Powder Technol.* **2009**, *194*, 239–245. [CrossRef]
40. Meng, X.; Zhu, J.; Zhang, H. The characteristics of particle charging and deposition during powder coating processes with ultrafine powder. *J. Phys. D Appl. Phys.* **2009**, *42*, 12. [CrossRef]
41. Parodi, E.; Govaert, L.E.; Peters, G.W.M. Glass transition temperature versus structure of polyamide 6: A flash-DSC study. *Thermochim. Acta* **2017**, *657*, 110–122. [CrossRef]
42. Piazza, D.; Lorandi, N.P.; Pasqual, C.I.; Scienza, L.C.; Zattera, A.J. Influence of a microcomposite and a nanocomposite on the properties of an epoxy-based powder coating. *Mater. Sci. Eng. A* **2011**, *528*, 6769–6775. [CrossRef]
43. Mikhailov, M.; Yuryev, S.; Lapin, A.; Lovitskiy, A. The effects of heating on $BaSO_4$ powders' diffuse reflectance spectra and radiation stability. *Dye. Pigment.* **2019**, *163*, 420–424. [CrossRef]
44. Moussa, O.; Vassilopoulos, A.P.; Keller, T. Experimental DSC-based method to determine glass transition temperature during curing of structural adhesives. *Constr. Build. Mater.* **2012**, *28*, 263–268. [CrossRef]
45. Moussa, O.; Vassilopoulos, A.P.; Keller, T. Effects of low-temperature curing on physical behavior of cold-curing epoxy adhesives in bridge construction. *Int. J. Adhes. Adhes.* **2012**, *32*, 15–22. [CrossRef]
46. Bajat, J.B.; Popic, J.P.; Miskovic-Stankovic, V.B. The influence of aluminum surface pretreatment on the corrosion stability and adhesion of powder polyester coating. *Prog. Organ. Coat.* **2010**, *69*, 316–321. [CrossRef]

47. Bhargava, S.; Lewis, R.D.; Kubota, M.; Li, X.; Advani, S.G.; Deitzel, J.M.; Prasad, A.K. Adhesion study of high reflectivity water-based coatings. *Int. J. Adhes. Adhes.* **2013**, *40*, 120–128. [CrossRef]
48. Strzelec, K.; Pospiech, P. Improvement of mechanical properties and electrical conductivity of polythiourethand- modified epoxy coatings filled with aluminum powder. *Prog. Org. Coat.* **2008**, *63*, 133–138. [CrossRef]

 *processes*

Article

Experimental Studies on Corrosion Behavior of Ceramic Surface Coating using Different Deposition Techniques on 6082-T6 Aluminum Alloy

Ali Algahtani [1,2], Essam R.I. Mahmoud [3,4,*], Sohaib Z. Khan [3] and Vineet Tirth [1]

[1] Department of Mechanical Engineering, College of Engineering, King Khalid University, Abha 61413, Saudi Arabia; alialgahtani@kku.edu.sa (A.A.); vtirth@kku.edu.sa (V.T.)
[2] Research Center for Advanced Materials Science (RCAMS), King Khalid University, Abha 61413, Saudi Arabia
[3] Department of Mechanical Engineering, Faculty of Engineering, Islamic University of Madinah, Medina 41411, Saudi Arabia; szkhan@iu.edu.sa
[4] Central Metallurgical Research and Development Institute (CMRDI), Cairo 11421, Egypt
* Correspondence: emahoud@iu.edu.sa; Tel.: +966-56-865-3429

Received: 24 October 2018; Accepted: 21 November 2018; Published: 26 November 2018

Abstract: Aluminum alloys cannot be used in aggressive corrosion environments application. In this paper, three different surface coating technologies were used to coat the 6082-T6 aluminum alloy to increase the corrosion resistance, namely Plasma Electrolytic Oxidation (PEO), Plasma Spray Ceramic (PSC) and Hard Anodizing (HA). The cross-sectional microstructure analysis revealed that HA coating was less uniform compared to other coatings. PEO coating was well adhered to the substrate despite the thinnest layer among all three coatings, while the PSC coating has an additional loose layer between the coat and the substrate. X-ray diffraction (XRD) analysis revealed crystalline alumina phases in PEO and PSC coatings while no phase was detected in HA other than an aluminum element. A series of electrochemistry experiments were used to evaluate the corrosion performances of these three types of coatings. Generally, all three-coated aluminum showed better corrosion performances. PEO coating has no charge transfer under all Inductive Coupled Plasma (ICP) tests, while small amounts of Al^{3+} were released for both HA and PSC coatings at 80 °C. The PEO coating showed the lowest corrosion current density followed by HA and then PSC coatings. The impedance resistance decreased as the immersion time increased, which indicated that this is due to the degradation and deterioration of the protective coatings. The results indicate that the PEO coating can offer the most effective protection to the aluminum substrate as it has the highest enhancement factor under electrochemistry tests compared to the other two coatings.

Keywords: 6082-T6 aluminum alloy; Plasma Electrolytic Oxidation (PEO); Plasma Spray Ceramic (PSC); Hard Anodizing (HA); anodic polarization; corrosion resistance

1. Introduction

Aluminum alloy 6082 has the highest strength of the 6000 series alloys and it used in many highly stressed applications such as aeronautics, trusses, bridges, transport applications, cranes and aerospace industries [1]. Aluminum is a metal which has a natural corrosion resistance due to the oxide layer that forms on its surface [1]. This dense layer is formed in a short time when it is exposed to the environment. However, under aggressive environments, aluminum is subjected to different types of corrosion such as pitting corrosion [2], intergranular corrosion [3] and stress corrosion cracking [4]. Pitting corrosion usually attacks aluminum surfaces causing localized holes in the protective film under chloride corrosive environments [5]. Regarding the atmospheric corrosion of aluminum,

El-Mahdy et al. [6] studied the effects of aluminum corrosion rate under cyclic wet-dry environments. They found that aluminum corrosion rate can be affected by temperature, surface inclination and relative humidity. The corrosion rate increases with increasing temperature and decreases by increasing the angle of inclination [6]. Corrosion plays a significant role in human life and safety. Corrosion attacks the component and affects its function negatively and consequently, reduces its service lifetime. The economic costs of corrosion are obviously enormous in many industrial fields [7,8]. As a result, relatively poor corrosion resistance often decreases the lifetime of the aluminum alloy components. Thus, some surface engineering techniques on aluminum alloys would be indispensable to their applications.

Extensive researches on surface engineering that enhances the material resistance against corrosion, specifically, started in the 1980s and became a recognized manufacturing technology in the 1990s [9]. There are two main advantages of using surface treatments. Firstly, surface engineering techniques can enhance the tribological performances of the component's surfaces against the surrounding environments and consequently, increase service lifetime and reduce the cost of replacement and maintenance [10]. Secondly, they can give a wide range of options for selecting cheaper substrate materials for certain applications with surface engineering technology [11,12].

Surface engineering technology can be classified based on the change of the surface of the substrate [13]. The first group can enhance the surface without changing the chemical composition of the substrate using heat treatments or mechanical working. Enhancing the surface by the second group through changing the chemical composition of the substrate such as electroplating and thermal diffusion treatments, oxide coatings, anodizing and sulphur treatments. The third group can enhance the surface by adding a layer such as welding, laser alloying and thermal spraying [14–16].

Plasma Electrolytic Oxidation (PEO) is an electrochemical surface treatment that produces a protective surface coating on metals and alloys [17–19]. This surface coating, which is ceramic, is produced by passing a modulated electrical current through a path of the electrolyte solution. The applied electrical potential should be high enough to the plasma discharges and sparks to be formed and generate oxide films with relatively higher thickness [20–22]. The applied potential exceeds the breakdown strength of the growing oxide film. The coated surface produced is well adhered to the substrate and possesses good wear and corrosion resistance, in addition to a good surface barrier for thermal conductivity [23,24].

The Hard Anodizing (HA) technique is an electrolytic passivation process, which forms a thick oxide layer on the metal surface. It is considered to be a traditional coating method and it has been used since 1923 and commercially available since 1940 [25–27]. HA is relatively a cheap process and does not give good wear resistance due to low hardness. This process can give better surface corrosion resistance, which varies according to the substrate material, parameters of the process and treatments conditions [28]. The microstructure of the anodized aluminum is considered amorphous alumina. This porous layer slightly increases the corrosion resistance of the substrate, however, for significant corrosion protection, a sealing process is required [29,30].

Plasma Spray Ceramics (PSC) is used to produce a coating in which molten or softened particles are applied by impacting onto a substrate. It was first introduced by Schoop while studying the production of metallic particles from a molten metal for coating [31–34]. There are three main stages to form this coating. Firstly, plasma particles are created as small droplets stream [35]. Secondly, these particles are subjected to a high temperature using heat source generating thermal energy. The particles' composition is changed due to the chemical reaction between the droplet material and the flame. After that, they are flattened while striking a cold surface at high velocities. A common feature of lamellar grain microstructure is formed as a result of the rapid solidification and cooling processes [36].

The characteristics of the PEO coating interface on AA1060 aluminum alloy were investigated by Wang et al., as a function of PEO processing time using scanning electron microscopy and X-ray diffraction. Hundreds of coatings were detached from the substrate by an electrochemical method and ground into homogeneous powders [37]. Shifeng et al. [38] investigated the morphology and protective properties of the PEO coatings produced on 5754 aluminum alloy in a mixed electrolyte. The current density on the sample surface increased, the coating grew faster, the thickness increased and the roughness gradually decreased. The coating corrosion resistance first increased and then decreased [38]. Cerchier et al. [39] used the Plasma Electrolytic Oxidation coating technique on samples of 7075 aluminum alloy and they obtained thick and adherent coatings [39]. Abdel-Gawad et al. [40] studied the corrosion behavior of the hard-anodizing coating formed on different groups of aluminum alloys in the sulphuric acid electrolyte. The corrosion resistance and the anodized coating layer are influenced by the type of the alloy and the anodizing conditions such as current density, acid concentration and time of duration [40]. The influence of the hard-anodizing process parameters such as H_2SO_4 concentration, electrolyte temperature, Al^{3+} concentration and current density for the AA2011-T3 was evaluated. Higher H_2SO_4 concentration and higher current density have improved coating hardness and defectiveness, however, potentiodynamic polarizations have revealed that they do not enhance corrosion resistance [41].

Many authors have evaluated the wear and corrosion behavior of surface coatings and most of them agreed that these coating materials increased the wear and corrosion resistance as compared to the uncoated ones [22,23,42,43]. The oxide film can be affected by changing treatment parameters depending on the purpose of the coating [44]. Also, it was found that as the thickness of the PEO coating increased, the corrosion resistance increased. For example, Qiu et al. [45] showed that the corrosion performances of the PEO coating on ZK60 Mg alloy could be improved by increasing the current density in the PEO process [45].

This work has investigated the enhancements of the three coatings; plasma electrolytic oxidation, plasma spray ceramic and hard anodizing on the performances of 6082 aluminum alloy surface against different corrosion experiments. The microstructure of the coating layer was detailed investigated and the comprehensive results are presented. In addition, the corrosion behavior of the coating layers was evaluated. The results are followed with the discussion of such behavior.

2. Materials and Methods

The substrate used in this study was 6082-T6 aluminum alloy with chemical composition listed in Table 1. T6 refers to the temper number which means solution heat-treated and artificially aged. The substrates were cut into discs with a diameter of 25.40 mm and a thickness of 10 ± 0.01 mm to be fitted in the holder for electrochemistry experiments suitable for the rig available in the lab. Three types of materials coatings were formed on the substrate, that is, Plasma Electrolytic Oxidation (PEO) and Plasma Spray Ceramic (PSC) and Hard Anodizing (HA). For PEO samples, a 3 mm diameter hole was drilled in the aluminum substrates for the anode to be inserted in the sample to allow for the anodization process and sent to Keronite International Ltd., Haverhill Suffolk, UK. For PSC samples, the aluminum substrate was sent to Bodycote Plc., Cheshire, UK where the Metco®101NS Grey Alumina Powder was used. The HA samples, coatings were provided by MP Eastern Ltd., Suffolk, UK.

Table 1. Chemical compositions of the 6082-T6 aluminum alloy; wt%.

Si	Mg	Mn	Fe	Cr	Cu	Zn	Ti	Al
0.957	0.836	0.634	0.55	0.25	0.12	0.24	0.13	Balance

A special sample holder for electrochemistry tests has been designed to allow testing the materials in a more convenient way as shown in Figure 1. This holder allows both faces of the sample to be tested without using resin. An Inductively Coupled Plasma (ICP) test was performed on all the coated samples together with the aluminum substrate to assess the number of ions (Al^{3+}) released to the 3.5% NaCl solution after 24 h of immersing the samples into the solution at different test conditions. Then, the samples were polarized up to 400 mV against the reference electrode for another 24 h. A series of electrochemistry experiments was used to evaluate the corrosion performances of three types of coatings deposited on 6082 aluminum alloy in the electrolyte of 3.5% NaCl solution. The reference electrode is sliver/sliver chloride (Ag/AgCl). The first experiment was Open Circuit Potential (OCP) tests where the potential (in Volts) was recorded against time (in seconds) for 24 h. The second experiment was direct current (DC) anodic polarization measurements which involve changing the electrode potential from its OCP in a certain direction and a given scan rate. Anodic Polarization (AP) tests are involved in measuring the scan from OCP to more positive voltages (up to 1 V from OCP) to reveal more information about the kinetics of the corrosion and its type. The determination of the corrosion current density i_{corr} will be done graphically from the plot of E. versus Log i, which is the intersection point of two lines. The third experiment was the AC impedance test which applied to the materials using an electrochemical measurement unit called Solarton (SI 1280B). The amplitude of the sinusoidal voltage was 10 mV which was selected to keep the system linear. The measurements were performed at frequencies ranging from the high value of 20 kHz to low-frequency value of 0.1 kHz to minimize the sample perturbation.

The microstructures of the coated layers and substrates were investigated using Leica optical microscope and Scanning electron microscope (Philips XL30 ESEM environmental SEM, Leeds, UK) equipped with Oxford Instruments INCA 250 EDX system analyzer after standard methods of metallography. The substrate and the coated layers were analyzed by X-ray diffractometer, (XRD, D8 Discover with GADDS system, 35 kV, 80mA, MoKα radiation, Leeds, UK) to identify the phases.

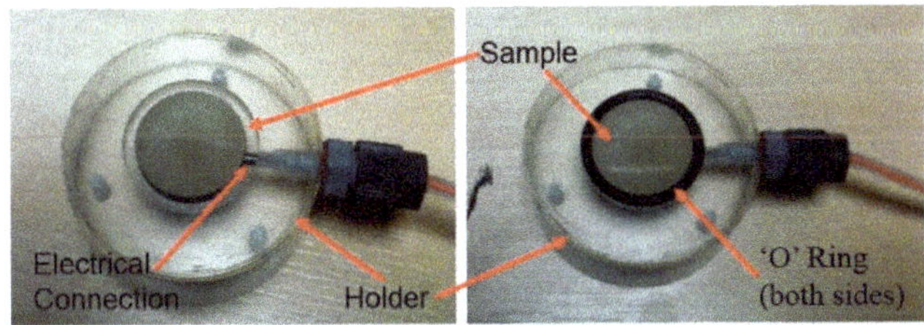

Figure 1. Designed sample holder for electrochemistry tests.

3. Results and Discussion

3.1. Macrol Microstructure Analysis

Figure 2 shows the microstructure of the substrate 6082-T6 aluminum alloy in polished and etched conditions. The micrographs revealed particles of different sizes distributed within the Al matrix. These particles may be the well-known intermetallics (β-Al$_5$FeSi, Mg$_2$Si, Al$_9$Mn$_3$Si, Mg$_2$Si, α-Al(FeMn)Si) shown in this type of Al alloy [46,47]. These intermetallics support and strengthen the matrix. For that, the alloy 6082 is considered one of the highest strength of the aluminum alloys.

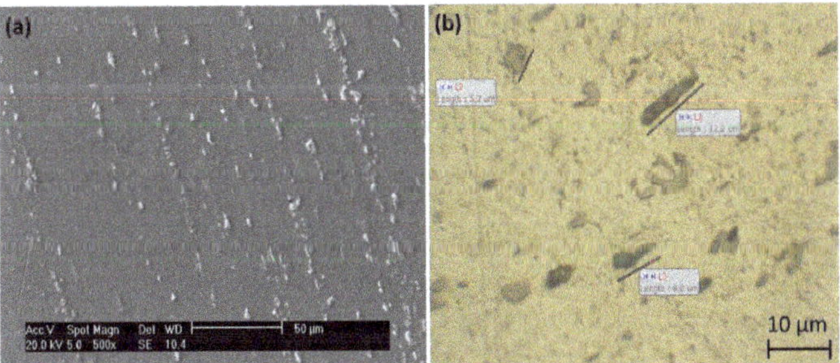

Figure 2. Microstructure of the substrate 6082-T6 aluminum alloy in (**a**) SEM polished micrograph and (**b**) Optical micrograph of the etched conditions.

The cross-sectional macrostructures of the coatings are shown in Figure 3. For all the three coatings, two main distinct layers can be seen in the cross-sectional view of the coating which is inner and outer layers. For HA (Figure 3a) the inner layer is dense and is composed of a thickness of 50 μm while the outer one is more porous with 10 μm in thickness. It can be shown that HA coating is less uniform and had more porosity compared with PEO coatings (Figure 3b). For PEO coating, the inner interface layer which is dense and well adhered to the substrate. This layer composes the major part of the coatings and it has a thickness of approximately 40 μm (Figure 3b). This interface coating layer has a uniform distribution with the substrate for both of them. From Figure 3c,d, PSC coating has the largest coating thickness of about 350 μm starting with a loose layer in the interface region of 85 μm. This layer is followed with the intermediate layer of approximately 250 μm and finally the top porous layer of 55 μm. The intermediate layer contains many laminar structures with white colors which could be the aluminum substrate. The dark layer (gap) at the interface is probably due to debonding of the PSC coating from the substrate. Although the deboning can be attributed to the metallographic preparation (sectioning/polishing); it also indicates a weaker bond with the substrate.

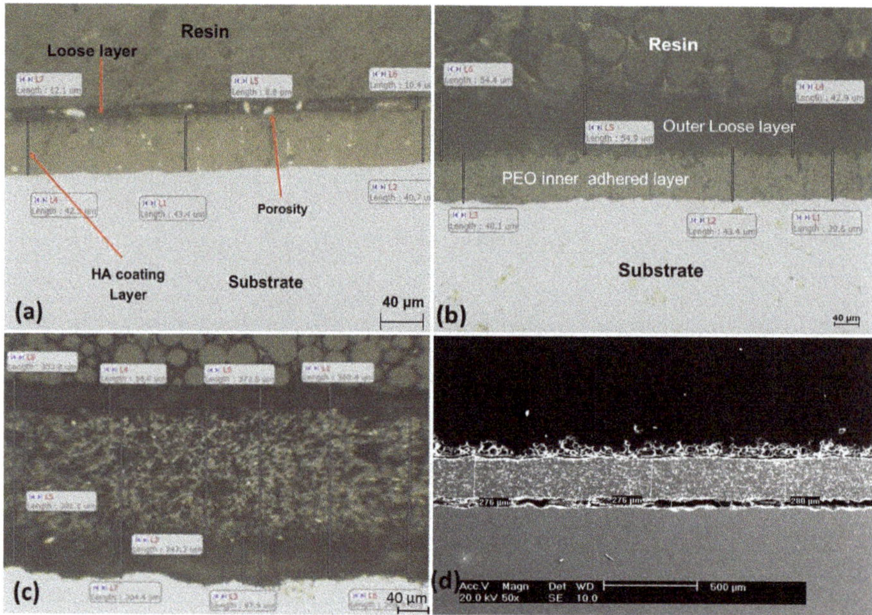

Figure 3. Optical cross-sectional micrographs of coating layers after polarization tests for (**a**) Hard Anodizing (HA), (**b**) Plasma Electrolytic Oxidation (PEO), (**c**) Plasma Spray Ceramic (PSC) and (**d**) SEM image of PSC.

The EDX spectra of hard anodizing reveal that the main element is oxygen followed with aluminum and sulphur which is due to the sulfuric acid (H_2SO_4) used in the electrolyte chemical composition during the HA process [48,49]. For PEO coatings, the main elements are aluminum, oxygen, manganese and magnesium as detected from the EDX analysis. These elements are due to the inclusion of the aluminum alloy elements. In addition, the EDX spectra showed titanium for PSC coating. The PSC coating was prepared using grey alumina powder (PSC Metco®101NS, Cheshire, UK) which contains 2.5% of titanium dioxide according to the supplier [50].

For the XRD analysis, there are no phases detected for HA in XRD apart from aluminum element as shown in Figure 4 a,b. That is because the HA coating is amorphous in nature and is believed to be due to oxide hydration [51]. The main phases detected in the PEO coating are alumina phases

(α-Al$_2$O$_3$, γ-Al$_2$O$_3$) as is shown in Figure 4c. This result is consistent with the literature where α-Al$_2$O$_3$ was formed on the PEO applied on aluminum alloy on the inner layer of the coating due to the high temperature during the discharge stage of the PEO process. Also, the amorphous γ-Al$_2$O$_3$ alumina is abundant in the outer layer which is formed during the cooling stages because of the contact between the molten alumina and the electrolyte. Similar to PEO coatings, PSC coating has α-Al$_2$O$_3$ and γ-Al$_2$O$_3$ phases but with lower intensity peaks (Figure 4d).

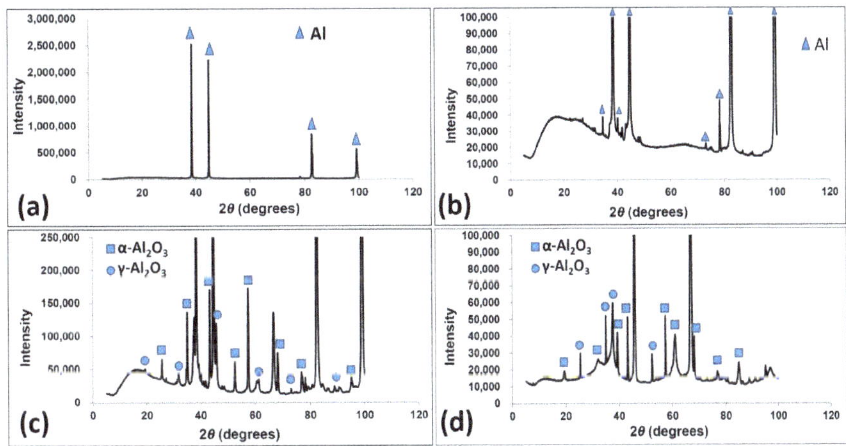

Figure 4. XRD spectra for coating layer of (**a**) HA, for highlighting lower peaks zoomed XRD spectra for (**b**) HA, (**c**) PEO and (**d**) PSC.

3.2. Electrochemistry

3.2.1. Inductive Coupled Plasma (ICP) Test

The amounts of aluminum ions (Al^{3+}) detected from the 3.5% NaCl solutions using ICP technique after each test condition for the three coated samples are presented in Table 2. It can be seen that no ions were found in the absence of any potential applied, except at high temperature for PSC sample only. After the 80 °C free corrosion tests, the samples were immersed in the solution for another 24 h and very small amounts of Al^{3+} were released for both HA and PSC coatings. However, PEO coating has no charge transfer under all test conditions.

Table 2. Al^{3+} (in grams) detected from 3.5% NaCl solution after 24 h free corrosion test.

Test Condition	HA	PEO	PSC
Free Corrosion at 20 °C	0.000	0.000	0.000
Free Corrosion at 80 °C	0.000	0.000	0.590
Free Corrosion at 20 °C after 80 °C	0.008	0.000	0.077

3.2.2. Mass Loss (Al^{3+}) from Polarization Tests

The amount of aluminum alloys released after polarizing the samples up to 400 mV from the OCP in 3.5% NaCl solution for 24 h are determined using the ICP method. Figure 5 shows the potential current versus time plots to calculate charge transfer at 400 mV for HA, PEO and PSC. There was a sharp increase in the current for the first 1200 s for HA then the current stabilized and steadily increased with some fluctuation in the rest of the test period. However, the PEO coating showed a gradual increase in current density throughout the test period without any sudden increase of the current which indicates a low rate of ion transfer through the coating. Also, the current values were an order of magnitude lower. There was no stability of the current in the case of the PSC sample as a high variation of the curve occurred during the 24 h polarization test and the current was significantly high compared to the other two coatings. The first increase of the current occurred at about 1000 s from 1.5×10^{-2} A/cm^2 to 3.0×10^{-2} A/cm^2 which could be attributed to high movements of Al^{3+} movement through the coating due to its high porosity. However, the current increased from about 2.5×10^{-2} A/cm^2 to 5.0×10^{-2} A/cm^2 around 7000 s where it is expected that the PSC coating has completely removed from the substrate. The current fluctuation on the polarization curves for all coatings could be attributed for electrochemical activities taking place on the surfaces such as localized pitting corrosion in the substrate/coating interface region.

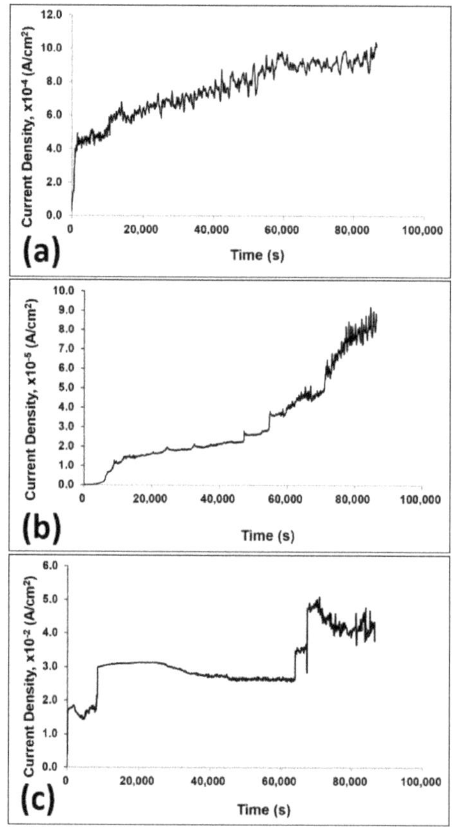

Figure 5. Current density versus time plots to calculate charge transfer for coating by (**a**) HA, (**b**) PEO and (**c**) PSC at +400 mV (Ag/AgCl).

3.2.3. Open Circuit Potential (OCP) Measurements

The Open Circuit Potential (OCP) measurements were carried out for the 30 s to find the starting OCP values for all the three coating. Many measurements were made and the average values are presented in Figure 6 for all the samples. This represented the initial OCP measurements when the samples were immersed in 3.5% NaCl without applying any potential. Firstly, it can be seen that the PSC sample has the most negative OCP value of about −0.76 V followed with the Al substrate with −0.69 V. It is expected that Al has a low negative value of E_{corr} since it can release three electrons per atom and consequently make the Al to be used as an anode in power sources applications.

Also, it has been reported that in chloride solutions it has an E_{corr} value of -0.75 V/SCE [52]. However, PEO coating has the highest starting OCP value with -0.048 nV while HA has around -0.4 V. According to [53], a high OCP value refers to better corrosion protection, which indicates the enhancements of these two coatings in the anticorrosion performances.

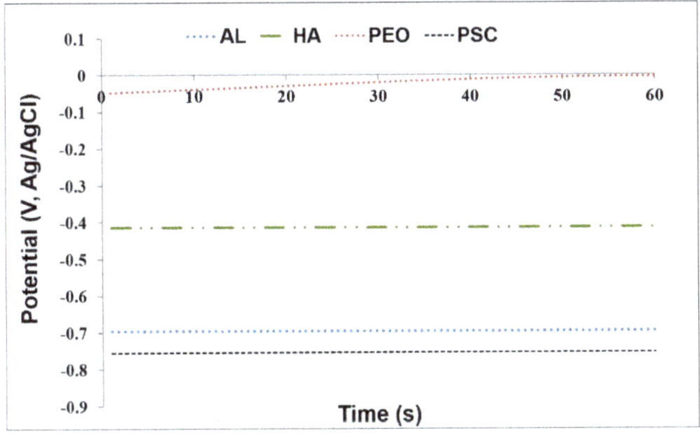

Figure 6. Initial Open Circuit Potential (OCP) measurements of the coating materials versus the Ag/AgCl reference electrode.

The OCP measurements have been extended for 5000 s, as shown in Figure 7 and it has been observed that the OCP value for the PEO sample dropped to the OCP value for the substrate after about 650 s. This sudden change in the OCP curve can indicate a rapid movement of ions through the coating part (insulator). However, the HA coating continued until approximately 3350 s then it decreased sharply to the OCP value for Al. The rapid decrease of the OCP value for PEO coating compared with the HA coating could be due to the difference in coating thickness where HA (42 µm) is higher than PEO (34 µm) which increase the resistance of Al^{3+} ions to penetrate through the coating [54]. However, the OCP value of the PSC coating keeps constant at a lower value than the OCP of the aluminum throughout the time of immersion.

In general, anodic areas are, at least in the early stages, much smaller than cathodic areas. So, in the early stages, the corrosion potential is more positive. However, with an increase in anodic sites during immersion, the corrosion potential becomes more negative.

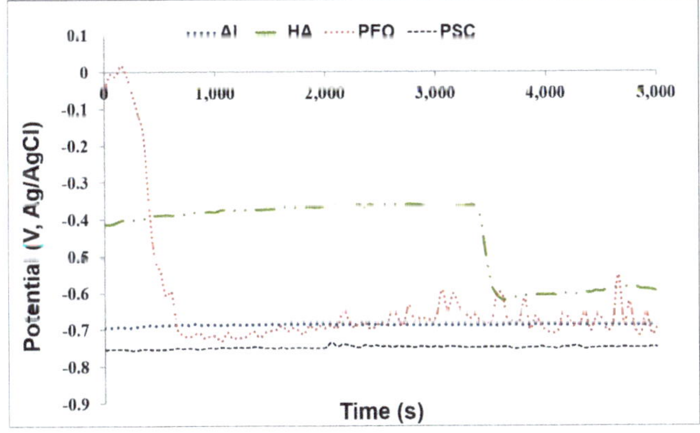

Figure 7. OCP measurements of the coating materials for 5000 s in Ag/AgCl reference electrode.

3.2.4. Anodic Polarization (AP) Resistance

The results of the AP resistance for all materials Al, HA, PEO and PSC samples are shown in Figure 8. The breakdown potential (E_b) is the potential where the passive film of the surface breaks down. It is expected that the coatings would decrease the possibility of breaking down the aluminum passive film and consequently decrease the current density and improve the corrosion resistance. E_b can be obtained from the AP curve (E. vs. i) where the current density increases sharply after this point. If there is no crevice corrosion, E_b refers to the pitting corrosion. However, E_b for the ceramic materials indicates the penetration of the electrolyte ions through the coating defects to the substrate metal. As a result, material's failure can be formed as localized pitting corrosion. When the potential reaches a voltage of ± 1000 mV from OCP or reaching a given magnitude of current density (referred to as the threshold current), the potential decreased towards the OCP value. The breakdown voltages for all the materials systems can be determined from the anodic polarization curves at the potential value where the current increased rapidly and deviated from the initial growing rate. The values of the breakdown potentials of the materials were determined from the plots and the red lines in these graphs are just to show the method and not indicating the exact E_b values. These values are summarized in Figure 9.

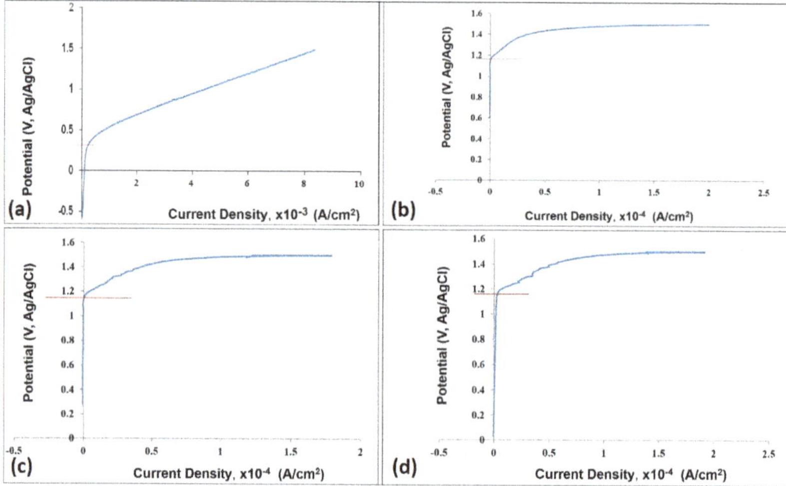

Figure 8. AP Measurements for (**a**) Al substrate, (**b**) HA, (**c**) PEO and (**d**) PSC coatings.

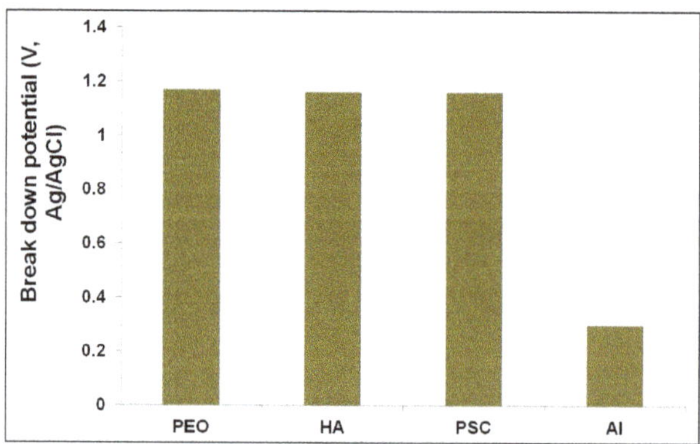

Figure 9. Comparison of breakdown potential from AP measurements for the tested materials against the Ag/AgCl reference electrode.

3.2.5. Corrosion Current Density

The corrosion current densities for the materials tested in this study were determined from the logarithmic scale of the current density in the anodic polarization curves as shown in Figure 10. The potential was shifted from the OCP value of the material to 250 mV in the opposite direction to ensure that the cathodic and anodic currents are different to measure the corrosion current density on the sample by extrapolating the anodic branch line from OCP value. Also, a comparison of the corrosion current density (i_{corr}) between all the coated materials is shown in Figure 11. The result of Al substrate was not included due to the huge difference values of i_{corr} between the aluminum substrate and other coatings. It is clear form this comparison chart that the PEO coating showed the lowest corrosion current density followed by HA and then PSC coatings.

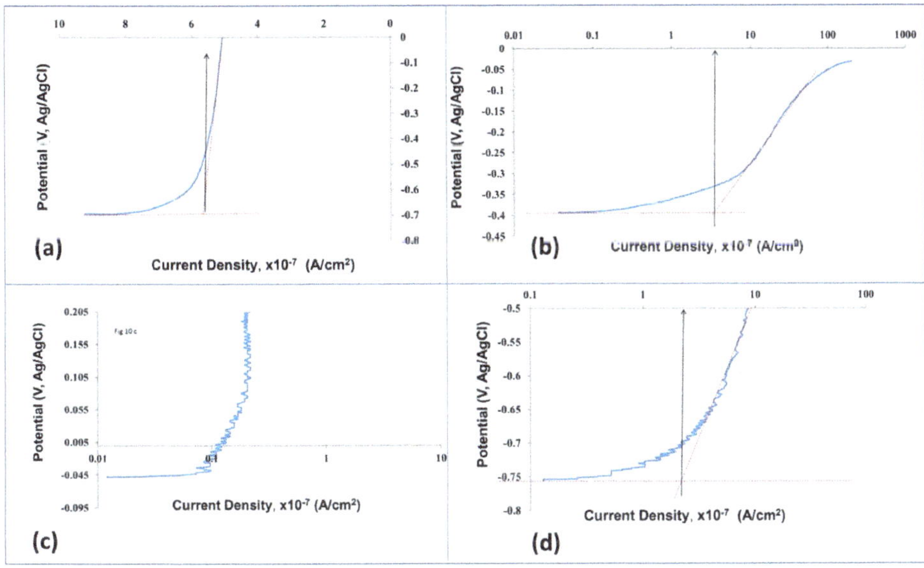

Figure 10. Determination of corrosion current density for (**a**) Al substrate, (**b**) HA, (**c**) PEO and (**d**) PSC coatings.

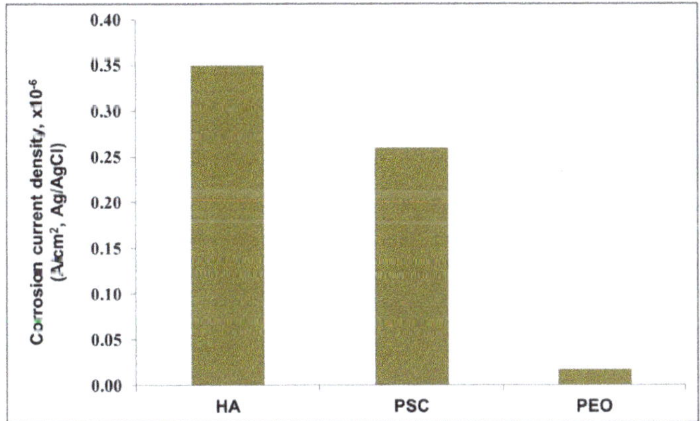

Figure 11. Comparison of corrosion current density values from AP measurements for the coated test specimens.

3.2.6. Optical Images of the Surfaces after Polarization tests

Figure 12 shows the surface behavior of the tested materials after Polarization scan in 3.5% NaCl solution. The aluminum surface (Figure 12a) has wide corrosion attack in the form of pits (see arrows). However, HA and PEO samples (Figure 12b,c, respectively) have fewer corrosion defects on the exposed surfaces. Regarding the PSC sample, a significant number of white spots (Figure 12d) on its surface can be seen which correspond to the aluminum substrate. From all these figures, the aluminum surface had more corrosion products as large-scale pits were initiated on its surface after the polarization test.

Table 3 summarizes the main corrosion parameters of the materials that were determined by the DC electrochemistry plots (anodic polarization curves). Therefore, PEO coating has lower corrosion current density (1.7×10^{-8} A/cm^2) than the HA coating (3.5×10^{-7} A/cm^2) and PSC coating (2.6×10^{-7} A/cm^2) under static anodic polarization tests.

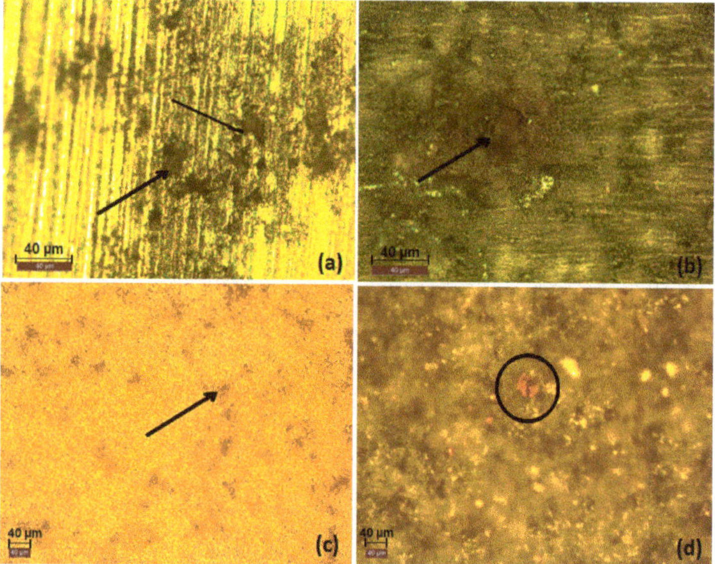

Figure 12. Optical images of coating surfaces after polarization tests for (**a**) Al, (**b**) HA, (**c**) PEO and (**d**) PSC coatings.

Table 3. Summary of corrosion parameters of the materials.

Corrosion Parameter	PEO	HA	PSC	Al
Open circuit potential (OCP) (V)	-0.05 ± 0.00246	-0.40 ± 0.01375	-0.76 ± 0.05542	-0.7 ± 0.005
Breaking down potential (E_b) (V)	1.07 ± 0.1	1.11 ± 0.05	1.12 ± 0.04	0.2 ± 0.1
Corrosion current density (i_{corr}) (A/cm^2) from AP Curve	1.7×10^{-8}	3.5×10^{-7}	2.6×10^{-7}	1.2×10^{-5}

3.2.7. AC Impedance Test

The Nyquist plots for all materials are presented (Figures 13 and 14) for 10 days of immersion the samples in 3.5% NaCl solution to study the stability of the materials (aluminum passive film of the substrate and coating part of the other materials) and observe the change in the total resistance during the long exposure period. Also, the impedance data of the materials were fitted with equivalent circuit models using ZView software (Ametrek, NC, USA).

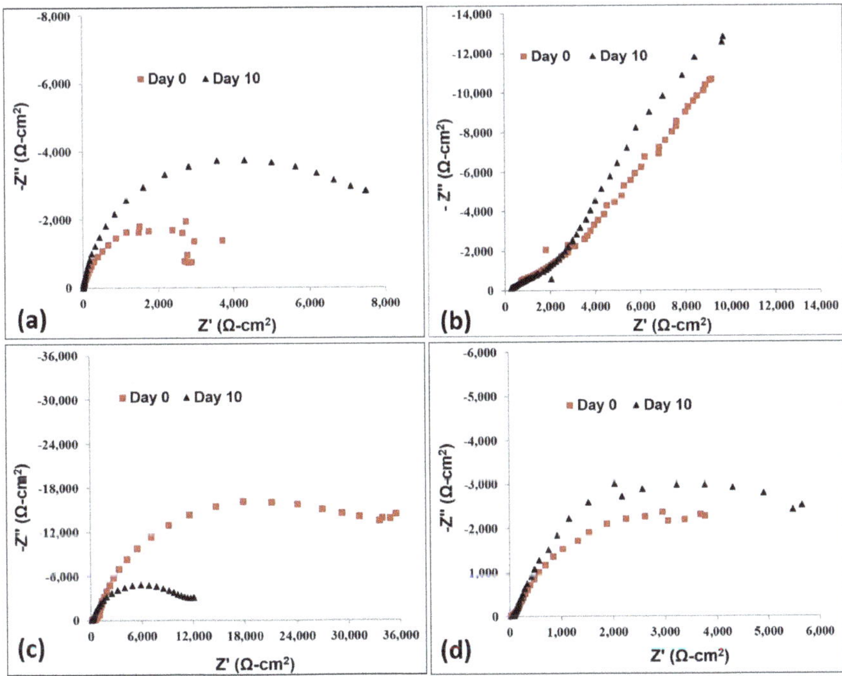

Figure 13. Nyquist plots for (**a**) Al substrate, (**b**) HA, (**c**) PEO and (**d**) PSC coatings at different immersion times.

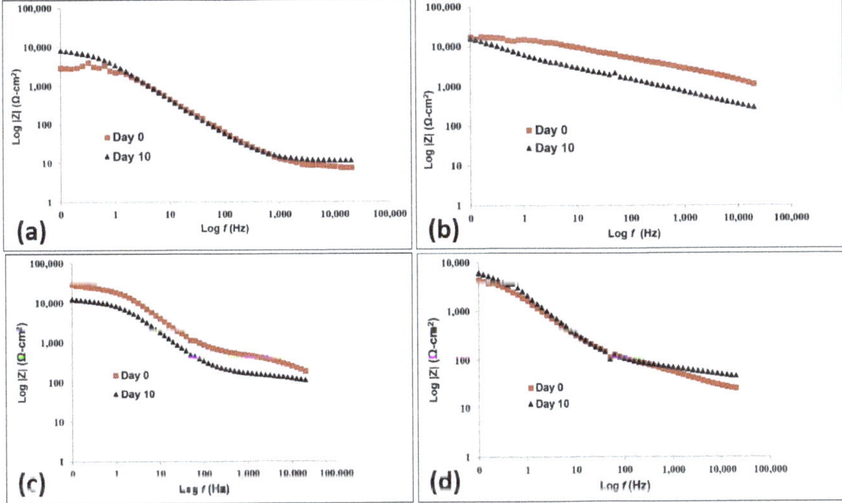

Figure 14. Bode plots for (**a**) Al substrate, (**b**) HA, (**c**) PEO and (**d**) PSC coatings at different immersion times.

The impedance of all spectra from Day 0 and Day 10 of the Al substrate exhibits a capacitive behavior pattern (single semicircle) as shown in Figure 13a. At initial immersion, the aluminum has the highest value of the total resistance (R_{tot}) which is corresponding to the large difference in the imaginary impedance after 24 h of immersion. The charge transfer resistance seems to increase as the exposure time increases and the maximum R_{tot} (maximum radius) is observed at day 10 (Figure 13a). Also, the highest peak in the Bode diagram is recorded on the last day as in Figure 14a. Therefore, it is consistent with the increase in the radius in Nyquist plots over the entire period. Many cavities with different sizes, some of them can be seen by eyes (Figure 15a), were formed on the passive aluminum surface due to the contact of the corrosive aqueous media (3.5% NaCl). Although the mechanism of the pitting corrosion is not fully understood, it can be explained by two consequent stages. Firstly, the pits are developed due to the adsorption of chloride ions Cl^- on the oxide film which cracks at weak areas causing micro-cracks and these pits get re-passivated as in Figure 15b. The intermetallic phases under the oxide layer have a low oxygen level driving the aluminum to by highly oxidized at the film broken sites. Secondly, the pits propagate due to the oxidation at the anode site (inside the pit) and the reduction at the cathode suite (outside the pits).

The AC Nyquist spectra for HA for all the period exhibited capacitive behavior (Figure 13b). Also, the stability of the capacitive behavior of HA is related to the high corrosion resistance of the HA coating. However, the slight decrease of the capacitive response at Day 10 (Figure 14b) could indicate the dissolution of the coating thickness or high porous coating was formed to allow electrolyte ions movements through the coated part [54]. For the surface behavior of HA sample after AC impedance test, different sizes of pits have been initiated on HA surface after 10 days of immersion the sample in 3.5% NaCl as shown in Figure 16. Similar pits were found in an anodized aluminum surface after a polarization test in 0.5 M NaCl solution performed by Ren et al. [55], which indicates penetration defect in the anodic film.

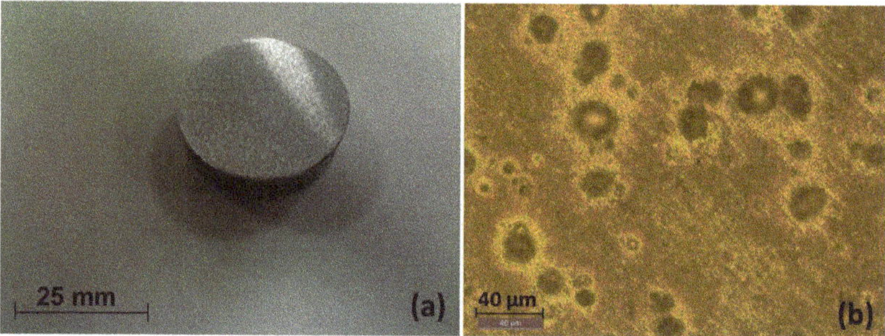

Figure 15. Optical images of aluminum surface after 10 days of immersion in 3.5% NaCl solution at (**a**) low magnification, (**b**) enlarged magnification.

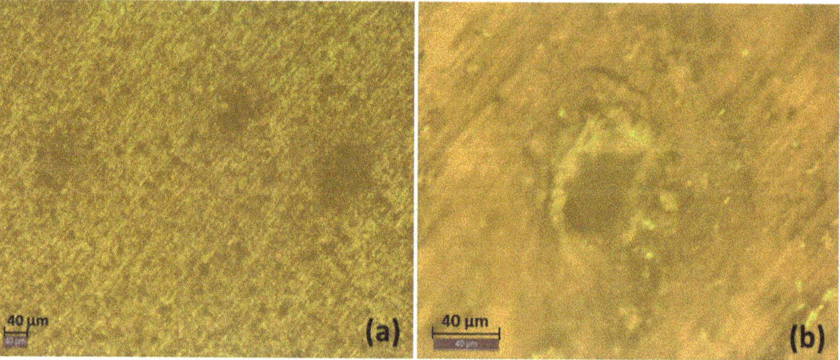

Figure 16. Optical images of HA coating surface after 10 days of immersion in 3.5% NaCl solution (**a**) low magnification, (**b**) enlarged magnification.

Regarding the AC impedance of the PEO coating, as shown in Figures 13 and 14, there is an arc at the high frequency side with a change to a diffusion tail at the low frequency side. The starting peaks of the impedance in the Bode magnitude plot is the highest for PEO sample in the initial immersion time. The drop of the starting peak after 10 days of the immersion (Figure 14c) indicates a change in the corrosion process taking place in the interface layer between the Al substrate and PEO coating during this period. The surface behavior of PEO coating after AC tests is shown in Figure 17. General corrosion products on the PEO surface is shown in Figure 17a while some materials degradation take place as shown in the magnified image in Figure 17b.

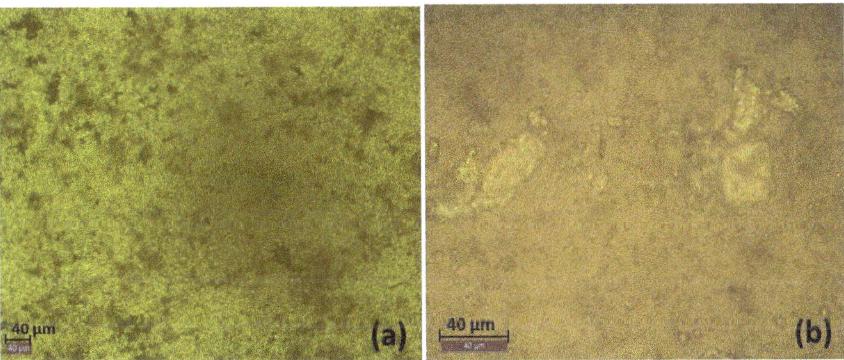

Figure 17. Optical images of PEO coating surface after 10 days of immersion in 3.5% NaCl solution (**a**) low magnification, (**b**) enlarged magnification.

The arc radius in the Nyquist plot of PSC coating was slightly increased after the full period of immersion test (Figure 13d). This corresponds to the minimal change in the total resistance from the Bode plot shown in Figure 14d. The aluminum substrate can be seen after the corrosion test (Figure 18) which indicates that a relatively high amount of PSC coating degradation has occurred due to this chemical reaction.

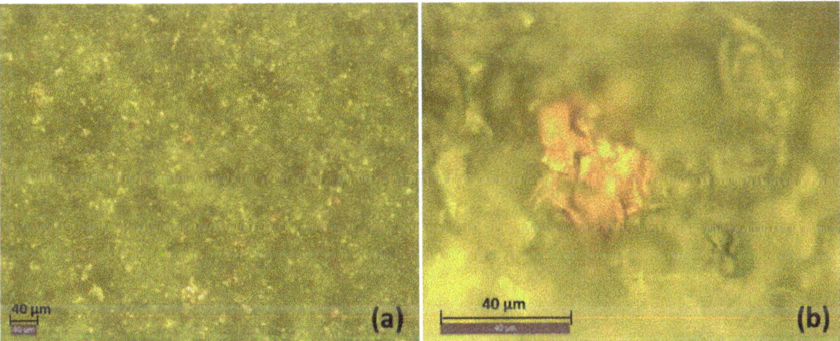

Figure 18. Optical images of PSC coating surface after 10 days of immersion in 3.5% NaCl solution (**a**) low magnification, (**b**) higher magnification.

The Nyquist and Bode curves for all tested materials have been combined in one plot to compare their AC impedance spectra. The Nyquist and Bode plots for all materials at the initial and after 10 days measurements are shown in Figures 19 and 20, respectively. It can be seen that both Al and PSC have depressed one semicircle, which means that only one-time constant has occurred in the system for both of them. This corresponds to one single peak in the Bode plot for Al and PSC. On the other hand, the PEO sample has two-time constants while HA has an infinite curve which is the behavior of a capacitive response. This capacitive form of HA indicates the stability of the coating passive film. After 10 days of immersion, the samples in the 3.5% NaCl solution, the Nyquist and Bode plots were obtained; and they are represented in Figures 19 and 20, respectively.

From the Bode magnitude plots, it can be seen that for all the material systems, the impedance resistance |Z| decreases as the immersion time increases; this indicates that material degradation is taking place in the protective nature of the coatings. This decrease makes the response less capacitive behavior which indicates that the solution ions penetrated through the pores in the coating to the substrate as the time of the immersion increased [54]. The sudden alteration of the |Z| versus frequency curve (Bode plot) is related to the coating degradation and significant changes in coating capacitance Cc and coating resistance Rc are noticed. The coating resistance slowly decreases when coating capacitance increases. This can be attributed to the higher porosity of the coating, which then creates heterogeneities in the coating and makes the water maneuver easier.

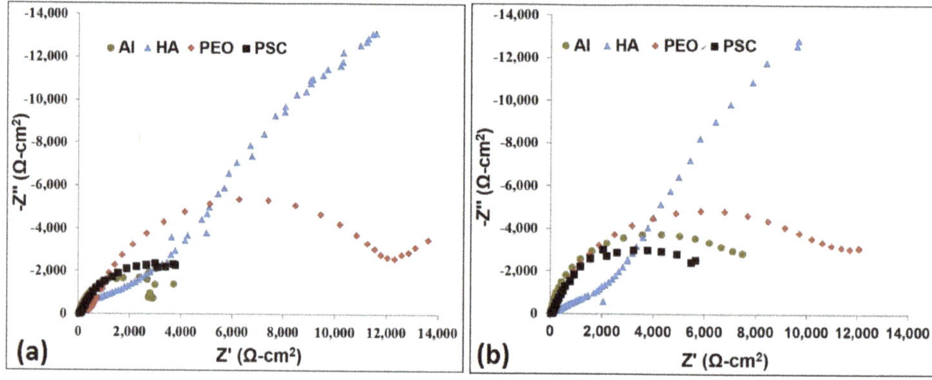

Figure 19. Nyquist plots for the materials after (**a**) 0 day and (**b**) 10 days.

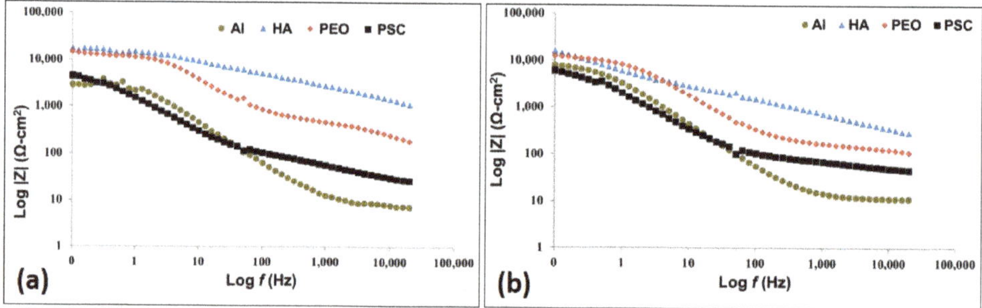

Figure 20. Bode plots for the materials after (**a**) 0 day and (**b**) 10 days.

4. Conclusions

This research has investigated the enhancements of different ceramic coatings (Plasma Electrolytic Oxidation (PEO), Plasma Spray Ceramic (PSC) and Hard Anodizing (HA)) on the performances of 6082-T6 aluminum alloy surface against corrosion environments. Coatings properties have been characterized using various metallurgical tools. A series of electrochemistry experiments were used to evaluate the corrosion performances of the three types of coatings. It has been shown that the corrosion performances of aluminum alloy can be highly increased using different surface treatments. PEO coating gives a higher level of corrosion resistance compared to the HA and the PSC coatings. The findings are summarized as follows:

- PEO coatings are denser, more uniform and well adhered to the substrate compared to the HA and the PSC coatings and show crystalline alumina phases (α-Al_2O_3, γ-Al_2O_3) structure.
- The PEO coating has the lowest amount of aluminum ions (Al^{3+}) released from the coating samples after immersion in 3.5% NaCl solution for 24 h followed with HA then PSC samples.
- The impedance decreases as the immersion time increases. It is concluded that this is due to the degradation and deterioration of the protective coatings.
- All the three coatings (PEO, HA and PSC) have better corrosion resistance than the aluminum substrate.

Author Contributions: Conceptualization, A.A.; Methodology, E.R.I.M. and S.Z.K.; Validation, V.T.; Formal Analysis, A.A., E.R.I.M. and S.Z.K.; Investigation, V.T.; Data Curation, A.A.; Writing—Original Draft Preparation, E.R.I.M. and A.A.; Writing—Review & Editing, E.R.I.M. and S.Z.K.; Supervision, A.A.; Project Administration, A.A.; Funding Acquisition, A.A.

Funding: This research was funded by [Scientific Research Deanship, King Khalid University (KKU)] grant number [G.R.P.2/6/38].

Acknowledgments: Authors thankfully acknowledge the funding and support provided by the Scientific Research Deanship, King Khalid University (KKU), Abha-Asir, Saudi Arabia, with grant number G.R.P.2/6/38 under the research group "Materials & Production" to complete the research work.

Conflicts of Interest: The authors declare no conflict of interest.

References

1. Mraied, H.; Cai, W.; Sagüés, A.A. Corrosion resistance of Al and Al–Mn thin films. *Thin Solid Film* **2016**, *615*, 391–401. [CrossRef]
2. Navaser, M.; Atapour, M. Effect of Friction Stir Processing on Pitting Corrosion and Intergranular Attack of 7075 Aluminum Alloy. *J. Mater. Sci. Technol.* **2017**, *33*, 155–165. [CrossRef]
3. De Bonfils-Lahovary, M.; Laffont, L.; Blanc, C. Characterization of intergranular corrosion defects in a 2024 T351 aluminum alloy. *Corros. Sci.* **2017**, *119*, 60–67. [CrossRef]
4. She, H.; Chu, W.; Shu, D.; Wang, J.; Sun, B. Effects of silicon content on microstructure and stress corrosion cracking resistance of 7050 aluminum alloy. *Trans. Nonferrous Met. Soc. China* **2014**, *24*, 2307–2313. [CrossRef]
5. Panagopoulos, C.N.; Georgiou, E.P. Corrosion and wear of 6082 aluminum alloy. *Tribol. Int.* **2009**, *42*, 886–889. [CrossRef]
6. El-Mahdy, G.A.; Kim, K.B. AC impedance study on the atmospheric corrosion of aluminum under periodic wet-dry conditions. *Electrochim. Acta* **2004**, *49*, 1937–1948. [CrossRef]
7. Nascimento, M.P.; Voorwald, H.J.C. Considerations on corrosion and weld repair effects on the fatigue strength of a steel structure critical to the flight-safety. *Int. J. Fatigue* **2010**, *32*, 1200–1209. [CrossRef]
8. Olajire, A.A. Corrosion inhibition of offshore oil and gas production facilities using organic compound inhibitors—A review. *J. Mol. Liq.* **2017**, *248*, 775–808. [CrossRef]
9. Wood, R.J.K. Tribo-corrosion of coatings: A review. *J. Phys. D Appl. Phys.* **2007**, *40*, 5502–5521. [CrossRef]
10. Ibatan, T.; Uddin, M.S.; Chowdhury, M.A.K. Recent development on surface texturing in enhancing tribological performances of bearing sliders. *Surf. Coat. Technol.* **2015**, *272*, 102–120. [CrossRef]
11. Kwok, C.T.; Man, H.C.; Cheng, F.T.; Lo, K.H. Developments in laser-based surface engineering processes: with particular reference to protection against cavitation erosion. *Surf. Coat. Technol.* **2016**, *291*, 189–204. [CrossRef]
12. Bebelis, S.; Bouzek, K.; Cornell, A.; Ferreira, M.G.S.; Walsh, F.C. Highlights during the development of electrochemical engineering. *Chem. Eng. Res. Des.* **2013**, *91*, 1998–2020. [CrossRef]
13. England, G. Surface Engineering in a Nutshell. Available online: http://www.surfaceengineer.co.uk/ (accessed on 10 November 2017).
14. Wang, C.; Bai, S.; Xiong, Y. Recent advances in surface and interface engineering for electrocatalysis. *Chin. J. Catal.* **2015**, *36*, 1476–1493. [CrossRef]
15. Zhang, X.; Chen, W. Review on corrosion-wear resistance performances of materials in molten aluminum and its alloys. *Trans. Nonferrous Met. Soc. China* **2015**, *25*, 1715–1731. [CrossRef]

16. Chi, Y.; Gu, G.; Yu, H.C.; Chen, C. Laser surface alloying on aluminum and its alloys: A review. *Opt. Lasers Eng.* **2018**, *100*, 23–37. [CrossRef]

17. Morks, M.F.; Corrigan, I.C.P.; Kobayashi, A. Electrochemical Characterization of Plasma Sprayed Alumina Coatings. *J. Surf. Eng. Mater. Adv. Technol.* **2011**, *1*, 107–111. [CrossRef]

18. Kossenko, A.; Zinigrad, M. A universal electrolyte for the plasma electrolytic oxidation of aluminum and magnesium alloys. *Mater. Des.* **2015**, *88*, 302–309. [CrossRef]

19. Martin, J.; Leone, P.; Nominé, A.; Veys-Renaux, D.; Belmonte, T. Influence of electrolyte ageing on the Plasma Electrolytic Oxidation of aluminum. *Surf. Coat. Technol.* **2015**, *269*, 36–46. [CrossRef]

20. Martin, J.; Melhem, A.; Ihchedrina, I.; Duchanoy, T.; Belmonte, T. Effects of electrical parameters on plasma electrolytic oxidation of aluminum. *Surf. Coat. Tech.* **2013**, *221*, 70–76. [CrossRef]

21. Kasalica, B.; Petkovic, M.; Belca, I.; Stojadinovic, S.; Zekovic, L. Electronic transitions during plasma electrolytic oxidation of aluminum. *Surf. Coat. Tech.* **2009**, *203*, 3000–3004. [CrossRef]

22. Simchen, F.; Sieber, M.; Lampke, T. Electrolyte influence on ignition of plasma electrolytic oxidation processes on light metals. *Surf. Coat. Tech.* **2017**, *315*, 205–213. [CrossRef]

23. Xie, H.; Cheng, Y.; Li, S.; Cao, J.; Cao, L. Wear and corrosion resistant coatings on surface of cast A356 aluminum alloy by plasma electrolytic oxidation in moderately concentrated aluminate electrolytes. *Trans. Nonferrous Met. Soc. China* **2017**, *27*, 336–351. [CrossRef]

24. Jiang, Y.; Zhang, Y.; Bao, Y.; Yang, K. Sliding wear behavior of plasma electrolytic oxidation coating on pure aluminum. *Wear* **2011**, *271*, 1667–1670. [CrossRef]

25. Massimiliano, B.; Manuela, C.; Roberto, G.; Andrea, B. Hard anodizing of AA2099-T8 aluminum-lithium-copper alloy: Influence of electric cycle, electrolytic bath composition and temperature. *Surf. Coat. Technol.* **2017**, *325*, 627–635.

26. Andrea, B.; Roberto, G.; Tiziano, M.; Paolo, M. Pulsed current effect on hard anodizing process of 7075-T6 aluminum alloy. *Surf. Coat. Technol.* **2016**, *270*, 139–144.

27. Massimiliano, B.; Roberto, G.; Andrea, B. Pulsed current hard anodizing of heat treated aluminum alloys: Frequency and current amplitude influence. *Surf. Coat. Technol.* **2016**, *307*, 861–870. [CrossRef]

28. Blawert, C.; Dietzel, W. Anodizing treatments for magnesium alloys and their effect on corrosion resistance in various environments. *Adv. Eng. Mater.* **2006**, *8*, 511–533. [CrossRef]

29. Huang, Y.; Shih, H. Evaluation of the corrosion resistance of anodized aluminum 6061 using electrochemical impedance spectroscopy (EIS). *Corros. Sci.* **2008**, *50*, 3569–3575. [CrossRef]

30. Wang, Y.; Jiang, S.L.; Zheng, Y.G.; Ke, W.; Wang, J.Q. Effect of porosity sealing treatments on the corrosion resistance of high-velocity oxy-fuel (HVOF)-sprayed Fe-based amorphous metallic coatings. *Surf. Coat. Tech.* **2011**, *206*, 1307–1318. [CrossRef]

31. Singh, H.; Sidhu, B.S. Use of plasma spray technology for deposition of high temperature oxidation/corrosion resistant coatings—A review. *Mater. Corros.* **2007**, *58*, 92–102. [CrossRef]

32. Kumar, S.; Kumar, A.; Kumar, D.; Jain, J. Thermally sprayed alumina and ceria-doped-alumina coatings on AZ91 Mg alloy. *Surf. Coat. Tech.* **2017**, *332*, 533–541. [CrossRef]

33. Thirumalaikumarasamy, D.; Shanmugam, K.; Balasubramanian, V. Corrosion performances of atmospheric plasma sprayed alumina coatings on AZ31B magnesium alloy under immersion environment. *J. Asian Ceram. Soc.* **2014**, *2*, 403–415. [CrossRef]

34. Sivakumar, S.; Praveen, K.; Shanmugavelayutham, G.; Yugeswaran, S. Thermo-physical behavior of atmospheric plasma sprayed high porosity lanthanum zirconate coatings. *Surf. Coat. Tech.* **2017**, *326*, 173–182. [CrossRef]

35. Li, C.; Wang, W. Quantitative characterization of lamellar microstructure of plasma-sprayed ceramic coatings through visualization of void distribution. *Mater. Sci. Eng. A* **2004**, *386*, 10–19. [CrossRef]

36. Li, C.J. Thermal spraying of light alloys. In *Surface Engineering of Light Alloys*, 1st ed.; Hanshan, D., Ed.; Woodhead Publishing: Cambridge, UK, 2010; pp. 184–241. ISBN 9781845699451.

37. Wang, R.Q.; Wu, Y.K.; Wu, G.R.; Chen, D.; He, D.L.; Li, D.; Guo, C.; Zhou, Y.; Shen, D.; Nash, P. An investigation about the evolution of microstructure and composition difference between two interfaces of plasma electrolytic oxidation coatings on Al. *J. Alloys Compd.* **2018**, *753*, 272–281. [CrossRef]

38. Liu, S.; Zeng, J. Effects of negative voltage on microstructure and corrosion resistance of red mud plasma electrolytic oxidation coatings. *Surf. Coat. Tech.* **2018**, *352*, 15–25. [CrossRef]

39. Cerchier, P.; Luca, P.; Emanuela, M.; Leonardo, C.; Marie, G.M.O.; Isabella, M.; Maurizio, M. Antifouling properties of different Plasma Electrolytic Oxidation coatings on 7075 aluminum alloy. *Int. Biodeterior. Biodegrad.* **2018**, *133*, 70–78. [CrossRef]

40. Abdel-Gawad, S.A.; Walid, M.O.; Amany, M.F. Characterization and Corrosion behavior of anodized Aluminum alloys for military industries applications in artificial seawater. *Surf. Interfaces* **2018**, in press. [CrossRef]

41. Massimiliano, B.; Roberto, G. Hard anodizing of AA2011-T3 Al-Cu-Pb-Bi free-cutting alloy: improvement of the process parameters. *Corros. Sci.* **2018**, *141*, 63–71.

42. Vatan, H.N.; EbrHAimi-kHArizsangi, R.; Kasiri-asgarani, M. Structural, tribological and electrochemical behavior of SiC nanocomposite oxide coatings fabricated by plasma electrolytic oxidation (PEO) on AZ31 magnesium alloy. *J. Alloys Compd.* **2016**, *683*, 241–255. [CrossRef]

43. Tjiang, F.; Ye, L.; Huang, Y.; Chou, C.; Tsai, D. Effect of processing parameters on soft regime behavior of plasma electrolytic oxidation of magnesium. *Ceram. Int.* **2017**, *43*, s567–s572. [CrossRef]

44. Hussein, R.O.; Northwood, D.O.; Nie, X. The effect of processing parameters and substrate composition on the corrosion resistance of plasma electrolytic oxidation (PEO) coated magnesium alloys. *Surf. Coat. Tech.* **2013**, *237*, 357–368. [CrossRef]

45. Qiu, Z.; Wang, R.; Wu, X.; Zhang, Y. Influences of Current Density on the Structure and Corrosion Resistance of Ceramic Coatings on ZK60 Mg Alloy by Plasma Electrolytic Oxidation. *Int. J. Electrochem. Sci.* **2013**, *8*, 1957–1965.

46. Majid, V.; Hyoung, S.K. A combination of severe plastic deformation and ageing phenomena in Al–Mg–Si Alloys. *Mater. Des.* **2012**, *36*, 735–740. [CrossRef]

47. Grażyna, M.N.; Sieniawski, J. Influence of heat treatment on the microstructure and mechanical properties of 6005 and 6082 aluminum alloys. *J. Mater. Process. Technol.* **2005**, *162*, 367–372. [CrossRef]

48. Malayoglu, U.; Tekin, K.C.; Malayoglu, U.; Shrestha, S. An investigation into the mechanical and tribological properties of plasma electrolytic oxidation and hard-anodized coatings on 6082 aluminum alloys. *Mater. Sci. Eng. A* **2011**, *528*, 7451–7460. [CrossRef]

49. Shrestha, S.; Shashkov, F. Microstructural and thermo-optical properties of black Keronite PEO coating on aluminum alloy AA7075 for spacecraft materials applications. In Proceedings of the 10th International Symposium, Materials in a Space Environment, Collioure, France, 19–23 June 2006.

50. Metco, S. Metco®101NS Grey Alumina Powder. Available online: https://www.oerlikon.com/ecomaXL/files/metco/oerlikon_DSMTS-0083.2_Al2O3_40TiO2.pdf&download=1 (accessed on 22 March 2017).

51. Krishna, L.R.; Purnima, A.S.; Sundararajan, G. A comparative study of tribological behavior of microarc oxidation and hard-anodized coatings. *Wear* **2006**, *261*, 1095–1101. [CrossRef]

52. Emregül, K.C.; Aksüt, A.A. The behavior of aluminum in alkaline media. *Corrosion Science* **2000**, *42*, 2051–2067. [CrossRef]

53. Wang, C.; Neville, A. Erosion-corrosion of engineering steels—Can it be managed by use of chemicals? *Wear* **2018**, *267*, 2018–2026. [CrossRef]

54. Cai, F.; Yang, Q.; Huang, X. The Roles of Diffusion Factors in Electrochemical Corrosion of TiN and CrN (CrSiCN) Coated Mild Steel and Stainless Steel. In *Materials Processing and Interfaces*; Wiley Online Library: Hoboken, NJ, USA, 2012; Volume 1. [CrossRef]

55. Ren, J.; Zuo, Y. The growth mechanism of pits in NaCl solution under anodic films on aluminum. *Surf. Coat. Technol.* **2005**, *191*, 311–316. [CrossRef]

 *processes*

Article

Model for Thin Layer Drying of Lemongrass (*Cymbopogon citratus*) by Hot Air

Thi Van Linh Nguyen [1,2,*], My Duyen Nguyen [1], Duy Chinh Nguyen [3], Long Giang Bach [3] and Tri Duc Lam [3]

[1] Faculty of Chemical Engineering and Food Technology, Nguyen Tat Thanh University, Ho Chi Minh City 70000, Vietnam; nguyenmyduyen155@gmail.com
[2] Centre of Excellence for Authenticity, Risk Assessment and Technology of Food, Nguyen Tat Thanh University, Ho Chi Minh City 70000, Vietnam
[3] NTT Hi-Tech Institute, Nguyen Tat Thanh University, Ho Chi Minh City 70000, Vietnam; ndchinh@ntt.edu.vn (D.C.N.); blgiang@ntt.edu.vn (L.G.B.); ltduc3096816@gmail.com (T.D.L.)
* Correspondence: ntvlinh@ntt.edu.vn

Received: 7 December 2018; Accepted: 28 December 2018; Published: 4 January 2019

Abstract: Lemongrass is a plant that contains aromatic compounds (myrcene and limonene), powerful deodorants, and antimicrobial compounds (citral and geraniol). Identifying a suitable drying model for the material is crucial for establishing an initial step for the development of dried products. Convection drying is a commonly used drying method that could extend the shelf life of the product. In this study, a suitable kinetic model for the drying process was determined by fitting moisture data corresponding to four different temperature levels: 50, 55, 60 and 65 °C. In addition, the effect of drying temperature on the moisture removal rate, the effective diffusion coefficient and activation energy were also estimated. The results showed that time for moisture removal increases proportionally with the air-drying temperature, and that the Weibull model is the most suitable model for describing the drying process. The effective diffusion coefficient ranges from 7.64×10^{-11} m^2/s to 1.48×10^{-10} m^2/s and the activation energy was 38.34 kJ/mol. The activation energy for lemongrass evaporation is relatively high, suggesting that more energy is needed to separate moisture from the material by drying.

Keywords: lemongrass; *Cymbopogon citratus*; convection drying; mathematical modeling; moisture diffusivity; activation energy

---

1. Introduction

Lemongrass (*Cymbopogon citratus*), also known as oil grass, silky heads, or citronella grass, is a popular annual crop for spices and medicinal herbs. The species is an herbaceous plant with long, thin leaves, and belongs to the Cymbopogon genus that is geographically distributed over various regions in India, America, Africa, Australia and Europe [1]. The use of lemongrass is diverse. At the family scale, lemongrass in both fresh and dried forms, is used as a spice in Vietnamese and Thai cooking, and for treatment of the common cold [2]. In Chinese and Indian medicine, the plant has been widely utilized as a tranquilizer and anti-inflammatory medicine. Tea made from lemongrass is also commonly consumed in Brazil, Cuba, and Argentina as a treatment for catarrh, rheumatism, and sore throat [3]. At larger scale, lemongrass is often cultivated and harvested to produce citronella oil and citronella-derived products such as mosquito repellant, soap, and perfume. The broad variety of uses of lemongrass is due to its valuable and numerous biological properties, which have been extensively studied and documented, including antifungal, antibacterial, antioxidant, anti-carcinogenic, and anti-rheumatic activities [3,4].

Lemongrass in general, and lemongrass oil in particular, is known for its abundant citral component, which is composed of geranial and neral. The content of citral in lemongrass essential oil varies from 40 to 82%, and is considered as an important quality indicator for utilization in medicine, food, herbicides, and cosmetics [5]. Lemongrass is also rich in aromatic compounds, which have been demonstrated to exhibit anticancer activity (myrcene and limonene) and pain-relieving properties (myrcene and limonene). Regarding antimicrobial properties, citral, mycrene, geranial, and gereniol—which are found in lemongrass in large proportions—have been shown to inhibit *Desulfovibrio alaskensis* (gram-negative bacteria), *Campylobacter jejuni* (enteritis-causing bacteria), *Escherichia coli* O157 (diarrhea-causing bacteria), and *Listeria monocytogenes* [6,7]. Other components with beneficial properties include limonene (antioxidant capacity), β-myrcene (gout prevention), citronellol (anti-fungal properties), and methyl heptentone (anti-diabetes, allergy, and anti-cancer) [6].

Similar to other aromatic plants, the nutritional quality of raw lemongrass cannot be maintained for an extended period of time, highlighting the need for the food industry to find a way to produce low-moisture lemongrass and select a suitable preservation method [8]. Drying, which involves the process of moving water inside the materials onto the surface and removing surface water by evaporation, is a common and efficient preservation technique [9]. Drying could prevent multiplication of spoilage-causing micro-organisms and minimize the occurrence of chemical reactions inside the material, in turn prolonging the shelf-life of the product and resulting in higher economic value than direct use of raw materials. Among many feasible methods of separating moisture from materials (such as convection drying, vacuum drying, freeze drying, and fluidized bed drying), thin layer hot air convection drying has proven to be a simple and cost-effective method [10]. However, to develop an efficient drying system for a particular product, the determination of suitable drying models for raw materials is required. Accurate modeling of material behavior during the drying process contributes to the development of dried products by predicting the drying process of materials based on the moisture balance of the material, from which improvements to existing drying system can be made [10].

Although lemongrass has been extensively studied with regards to its biological activities and chemical composition, literature concerning drying kinetics of lemongrass is still limited. To the best of our knowledge, two notable attempts have been made to model lemongrass behavior in drying processes [11,12]. The study of Coradi et al. (2014) suggested the appropriateness of the two-term model in describing the drying kinetics of lemongrass, demonstrated by a low value of the average estimated error (SE) and average relative error (P) [11]. However, only four typical models were taken into account in their study: modified Page, logarithmic, two-term and Midilli models. On the contrary, Simha (2016) fitted experimental data to an extensive pool of ten drying models, showing that the Midilli model is the most suitable drying kinetics model for both lemongrass and *Adathoda vasica* leaves [12]. However, the drying method of choice in this study was microwave drying, which lacks applicability and is energy-inefficient at larger scale.

Given the scarcity of studies on lemongrass drying kinetics and the significance of the thin layer convection-drying technique, this study aimed to determine the kinetic profile that characterizes convective drying of lemongrass material. In addition, the effective moisture diffusion coefficient, activation energy of the material, and other important parameters resulting from the present study can act as a precursor for further investigation regarding the mass transfer and moisture removal occurring during the drying process under various drying conditions.

2. Materials and Methods

2.1. Materials and Methods

2.1.1. Materials

Lemongrass (*Cymbopogon citratus*) was purchased at Vuon Lai Market, District 12, Ho Chi Minh City, Vietnam. Harvested lemongrass was planted within 10 to 12 months. The initial moisture content of the material was 6.40 ± 0.15 (g water/g dry matter). Before drying, lemongrass was pre-treated by washing and cut into uniformly sized samples with a length of 38.95 ± 1.35 cm and a thickness of 1.46 ± 0.14 mm (Figure 1). The moisture content was determined using the oven method [13].

Figure 1. The lemongrass material.

2.1.2. Drying Equipment

A pilot-scale hot air convection dryer was used to dry lemongrass slices. The dryer consisted of two drying rooms, in which each room contained 10 wire mesh trays. The drying tray size was 80 × 80 × 3 cm and the mesh size was 1 × 1 cm (Figure 2). Moist air from the external environment at 30.5 ± 1.0 °C and 70% RH was fed to the dryer by exhaust fans where it was heated to a maximum temperature of 65 °C. The heated air then came into contact with materials, absorbing moisture and exiting the dryer.

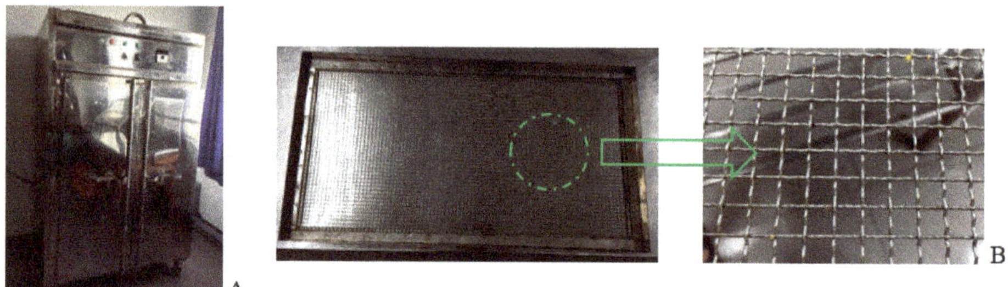

Figure 2. Convection dryer (**A**) and tray (**B**) used in this study.

2.1.3. Drying Procedure

For each sample, a thin layer (1000 g) was put in the tray dryer. In this study, the air-drying temperature was set at 50, 55, 60 and 65 °C. The sample weight was recorded during drying at 10-min interval. Experimental data was used to evaluate mathematical models of drying curve as well as calculate the effective moisture diffusivity and activation energy.

A drying temperature range of 50–65 °C is generally feasible for a wide number of medicinal plants [14] and was selected in this study for a number of reasons. First, it is suggested that, to obtain the highest essential oil content, drying of lemongrass should be conducted in an oven at 45 °C for 7 h [15]. In another study, it is suggested that temperature of drying should range from 50 to 70 °C to achieve maximum oil retention and that a temperature of below 30 °C was not advisable due to the promotion of fungi [16].Second, citral content, which determines the quality of the lemongrass products, was shown to be insensitive to thermal treatment in the range of 50–65 °C. Another study attempting drying of lemon myrtle leaves suggested that the optimal drying temperature for maximum citral retention was 50 °C [17] and that slightly higher drying temperature was not detrimental to citral content. This is due to the protective effects of the partially dried surface layers formed by high temperature, limiting the mass transfer and the diffusion of volatile components to the surface. This result is supported by Rocha et al., (2000) who stated that both the oil yield and composition of citronella (*Cymbopogon winterianus*) were at optimal levels at a drying temperature of 60 °C [18].

2.2. Determination of Mathematical Components

2.2.1. Moisture Content

Moisture content (g water/g dry matter) was calculated by the following equation [19].

$$M = \frac{m_w}{m_{dm}},$$

(1)

where M is the moisture content (g water/g dry matter), m_w is the mass of water in sample (g), and m_{dm} is the mass of dry matter in sample (g)

2.2.2. Drying Rate

Drying rate (DR) is defined as the amount of evaporated moisture over time. The drying rate (g water/g dry matter/min) during the process of drying lemongrass was determined using the following equation:

$$DR = \frac{M_t - M_{t+dt}}{dt},$$

(2)

where M_t is the moisture content at t time (g water/g dry matter), M_{t+dt} is the moisture content at t + dt time (g water/g dry matter), and dt is drying time (min).

2.2.3. Mathematical Modeling of Drying Curves

Moisture ratio is defined as follows [20].

$$MR = \frac{M_t - M_e}{M_0 - M_e},$$

(3)

where M_t is the moisture content (g water/g dry matter). The subscripts t, 0, and e denotes time t, initial, and equilibrium, respectively.

To identify the suitable mathematical model for lemongrass drying, the experimental data were fitted to different thin-layer drying models (Table 1).

Evaluation of mathematical models was performed based on coefficient of determination (R^2), root mean square error (RMSE) and chi-squared (x^2). The criteria for a suitable model included high a R^2 value, and low x^2 and RMSE values. The statistical values were defined as follows [21]:

$$\text{The coefficient of determination}: R^2 = 1 - \frac{\sum_{i=1}^{N} \left(MR_{exp,i} - MR_{pre,i} \right)^2}{\sum_{i=1}^{N} \left(\overline{MR_{exp,i}} - MR_{pre,i} \right)^2}$$

(4)

$$\text{The Root Mean Square Error}: RMSE = \sqrt{\frac{1}{N} \sum_{i=1}^{N} \left(MR_{exp,i} - MR_{pre,i} \right)^2}$$

(5)

$$\text{Chi} - \text{squared}: \chi^2 = \frac{\sum\limits_{i=1}^{N}\left(MR_{exp,i} - MR_{pre,i}\right)^2}{N - Z} \tag{6}$$

where MR_{exp} is the experimental dimensionless moisture ratio, MR_{pre} is the predicted dimensionless moisture ratio, $\overline{MR_{exp}}$ is the mean value of the experimental dimensionless moisture ratio, N is the number of observations, and Z is the number of constants in the mathematical model.

2.2.4. Estimation of the Effective Moisture Diffusivity

To determine the moisture diffusion of a material, the Fick diffusion model was applied as follows [21].

$$\frac{\partial M}{\partial t} = D_{eff}\frac{\partial^2 M}{\partial x^2}, \tag{7}$$

where M is the moisture content (g water/g dry matter), D_{eff} is the effective moisture diffusivity (m^2/s), and x is position with the dimensions of length (m).

Crank (1975) has provided a solution that deals with different shapes [22]:

$$MR = \frac{8}{\pi^2}\sum_{n=0}^{\infty}\frac{1}{(2n+1)^2}\exp\left(-(2n+1)^2\pi^2\frac{D_{eff}}{4L^2}t\right), \tag{8}$$

where MR is the dimensionless moisture ratio, L is the half-thickness (m), n is the term in series expansion, and t is time (s).

When drying for a long time, the above equation can be reduced to:

$$\ln(MR) = \ln\left(\frac{8}{\pi^2}\right) - \left(\pi^2\frac{D_{eff}}{4L^2}t\right). \tag{9}$$

The effective moisture diffusivity could be described by empirical data using the graph of ln(MR) versus time (t) and the slope of straight line from the plot as $-\pi^2\frac{D_{eff}}{4L^2}t$.

2.2.5. Estimation of Activation Energy

The activation energy is calculated by the Arrhenius equation [20,23]:

$$D_{eff} = D_0\exp\left(-\frac{E_a}{RT}\right), \tag{10}$$

where E_a is the activation energy (kJ/mol), R is the ideal gas constant (8.3143 kJ/mol), T is the absolute temperature (K), and D_0 is the pre-exponential factor (m^2/s).

2.2.6. Data Analysis

Microsoft Excel software was used to calculate moisture content, moisture ratio, and determine the effective moisture diffusivity and activation energy of the drying process. MATLAB 2014R software was used to find the best model fit and the mathematical model coefficients.

3. Results

3.1. Drying Curves and Drying Rate Curves

Figure 3 shows the change in moisture ratio of lemongrass at different temperatures, from the initial moisture content of 6.40 g water/g dry matter to the constant moisture content. It took 310, 290, 260 and 200 min to completely dry lemongrass at 50, 55, 60 and 65 °C, respectively. The results indicate that higher air-drying temperature resulted in greater slope of the curve and shorter drying time. The explanation for this could be two-fold. First, since moisture removal of the material occurs in parallel with moisture diffusion from center to surface material and from the surface to environment, higher air-drying temperature reduces relative humidity on the surface material, which in turn promotes surface evaporation in the drying process [23]. Second, increased air-drying temperature also leads to an improved temperature gradient and surface evaporation rate, accelerating moisture diffusion from the center to the surface. These results are consistent with another study, where the decreased drying time was attributed to increases of the air-drying temperature [24].

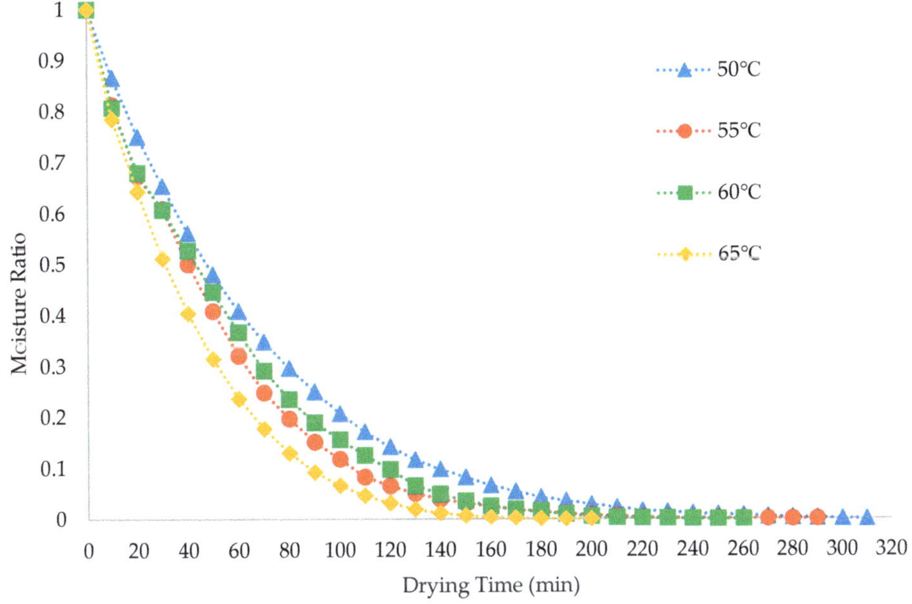

Figure 3. Drying curves of lemongrass slices at different temperature.

Figure 4 shows the drying rate curves of lemongrass slices at different temperatures. Evidently, higher drying temperature was associated with increased drying rate. Since higher temperature induces more heat transfer to the sample, which in turn leads to increased moisture diffusion to the inside and the outside of the materials, moisture removal was accelerated at higher air-drying temperatures. On the other hand, the drying process pattern described in Figure 4 only exhibited two periods—the initial and falling-rate periods. This is different from a standard drying process consisting of three periods: (i) initial period, (ii) constant-rate period, and (iii) falling-rate period. To be specific, the period starting from the initial moisture content to the moisture content of 5 g water/g dry matter coincided with the initial drying period and the period with moisture of higher than 5 g water/g dry matter represented the falling-rate period. The absence of the constant drying rate could be explained by the thin-layer arrangement and the high flow of the drying agent, which quickly accelerates evaporation and circumvents the saturation state of the material. Recent studies have also shown that the constant rate period was absent in the drying processes of fruits and vegetables, since this period often occurs very quickly [25]. At very low moisture content, of less than 0.2 g water/g dry matter, differences between drying rates of the four temperature levels were indistinguishable. This could be mainly due to the lack of water after the removal of free moisture, leading to the diversion of thermal energy into the breaking of bonds instead of heating water molecules. However, since the remaining water molecules were strongly bond to the cellulose fibers, increasing drying temperature from 50 to 65 °C was inadequate to induce a noticeable change in drying rate for lemongrass materials with low moisture.

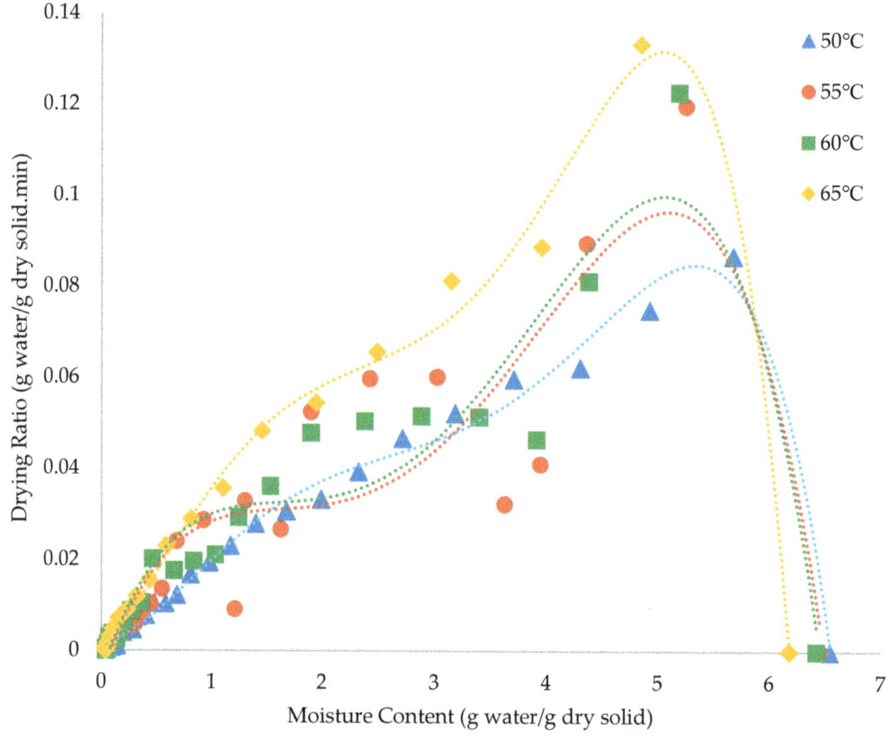

Figure 4. Drying rate curves of lemongrass slices at different temperatures.

3.2. Mathematical Models of Drying Curves

Seven common thin-layer drying models (Table 1) were fitted to assess the suitability of experimental data and the results of nonlinear regression analysis are presented in Table 2. The most suitable model to describe lemongrass drying is the model giving highest coefficient of determination (R^2), lowest Root Mean Square Error (RMSE), and lowest chi-square (x^2). From the results of Table 2, all models show that the R^2 values range from 0.93582 to 0.99983 and the RMSE values range from 0.00362 to 0.06887 and x^2 range from 1.50×10^{-5} to 5.08×10^{-5}. Therefore, any of these models of thin layer drying can be used to estimate the change in moisture content over time [26].

Table 1. Mathematical models used to predict the moisture ratios values [25].

No	Model	Equations
1	Page	$MR = \exp(-kt^n)$
2	Henderson and Pabis	$MR = a\exp(-kt)$
3	Midilli	$MR = a\exp(-kt^n) + bt$
4	Logarithmic	$MR = a\exp(-kt) + c$
5	Two-term	$MR = a\exp(-k_1 t) - b\exp(-k_2 t)$
6	Wang and Singh	$MR = 1 + at + bt^2$
7	Weibull	$MR = a - b\exp(-k_0 t^n)$

Table 2. Result of statistical analyses on the modeling of moisture content and drying time.

Temp.	Model	Coefficients				R^2	RMSE	x^2
50 °C		k = 0.00011	n = 1.10025			0.99959	0.00557	3.30×10^{-5}
55 °C	Page	k = 0.00015	n = 1.07613			0.99505	0.01912	3.92×10^{-4}
60 °C		k = 0.00014	n = 1.09420			0.99555	0.01861	3.74×10^{-4}
65 °C		k = 0.00017	n = 1.10771			0.99865	0.01048	1.21×10^{-4}
50 °C		a = 1.02592	k = 0.00026			0.99790	0.01266	1.71×10^{-4}
55 °C	Henderson	a = 1.00482	k = 0.00029			0.99377	0.02147	4.94×10^{-4}
60 °C	and Pabis	a = 1.00910	k = 0.00031			0.99361	0.02229	5.36×10^{-4}
65 °C		a = 1.01880	k = 0.00042			0.99661	0.01660	3.05×10^{-4}
50 °C		a = 0.99120	k = 0.00011	n = 1.09672	$b = -4.12 \times 10^{-7}$	0.99980	0.00386	1.65×10^{-5}
55 °C	Midilli	a = 0.96973	k = 0.00012	n = 1.09701	$b = -6.75 \times 10^{-7}$	0.99607	0.01703	3.22×10^{-4}
60 °C		a = 0.97110	k = 0.00012	n = 1.09938	$b = -1.19 \times 10^{-6}$	0.99704	0.01516	2.59×10^{-4}
65 °C		a = 0.98882	k = 0.00018	n = 1.09637	$b = -1.07 \times 10^{-6}$	0.99917	0.00821	7.86×10^{-5}
50 °C		a = 1.03448	k = 0.00025	c = −0.01985		0.99923	0.00769	6.52×10^{-5}
55 °C	Logarithm	a = 1.01527	k = 0.00027	c = −0.02295		0.99574	0.01775	3.50×10^{-4}
60 °C		a = 1.02510	k = 0.00028	c = −0.03240		0.99682	0.01573	2.70×10^{-4}
65 °C		a = 1.03218	k = 0.00038	c = −0.02517		0.99870	0.01027	1.23×10^{-4}
50 °C		a = 0.50040	$k_1 = 0.00026$	b = 0.49978	$k_2 = 0.00026$	0.99730	0.01435	2.35×10^{-4}
55 °C	Two-term	a = 0.55424	$k_1 = 0.00029$	b = 0.44483	$k_2 = 0.00029$	0.99373	0.02152	5.34×10^{-4}
60 °C		a = 0.50114	$k_1 = 0.00030$	b = 0.50058	$k_2 = 0.00031$	0.99356	0.02238	5.88×10^{-4}
65 °C		a = 0.49822	$k_1 = 0.00041$	b = 0.50202	$k_2 = 0.00041$	0.99628	0.01740	3.74×10^{-4}
50 °C		a = −0.00016	b = 0.00000			0.94964	0.06197	4.10×10^{-3}
55 °C	Wang and	a = −0.00017	b = 0.00000			0.93582	0.06887	5.08×10^{-3}
60 °C	Singh	a = −0.00019	b = 0.00000			0.95715	0.05773	3.60×10^{-3}
65 °C		a = −0.00025	b = 0.00000			0.95231	0.06229	4.29×10^{-3}
50 °C		a = −0.00803	b = −1.00008	$k_0 = 0.000117$	n = 1.08916	0.99983	0.00362	1.50×10^{-5}
55 °C	Weibull	a = −0.01311	b = −0.98494	$k_0 = 0.000138$	n = 1.08109	0.99620	0.01676	3.24×10^{-4}
60 °C		a = −0.02076	b = −0.99429	$k_0 = 0.000143$	n = 1.07915	0.99725	0.01463	2.51×10^{-4}
65 °C		a = −0.01382	b = −1.00375	$k_0 = 0.000199$	n = 1.08323	0.99927	0.00772	7.37×10^{-5}

Based on the results of the statistical analysis and the estimated coefficients of the mathematical models shown in Table 2, the Weibull model achieved the highest values of R^2, averaging at 0.99813, the lowest values of x^2, averaging at 0.000166, and lowest RMSE, averaging at 0.010683. This indicates that the model Weibull is the best model to estimate moisture ratio of lemongrass compared to the other models, followed by the Midilli model in which R^2, x^2, and RMSE averaged at 0.99802, 0.000169, and 0.011065 respectively. Since the Weibull model is rarely included in comparison studies of drying kinetics, the present results are in line with results of Onwude et al. (2016), demonstrating that approximately 24% of the literature sources supports the Midilli model; and with Simha et al. (2016), who affirmed the suitability of the Midilli model in microwave drying of *Cymbopogon citratus* [12,25]. From Table 2, it can also be seen that the predictive power of the Midilli model is weaker than that of the Weibull model in terms of R^2, RMSE, and x^2. At each air-drying temperature, the Weibull model exhibited higher R^2, and lower RMSE and x^2 in comparison with the other fitted models. In addition, the Weibull model estimates showed that the drying constant k_0 increased when the temperature increased from 50 to 65 °C. This result is consistent with previous studies, where the Weibull model was suggested to be able to well-describe the drying kinetics of fruits and various vegetables such as garlic, quinces, and persimmon [25].

3.3. Estimation of the Effective Moisture Diffusivity

Figure 5 shows the approximated linear relationship between ln(MR) and drying time (t) at different drying temperatures of 50, 55, 60 and 65 °C. Linear regression was used to calculate the effective moisture diffusion coefficient. The effective moisture diffusion coefficient and the R^2 correlation coefficient correspond to each temperature and were calculated by linear regression of experimental value. ln(MR) (dimensionless) with drying time (t) (second) are presented in Table 3.

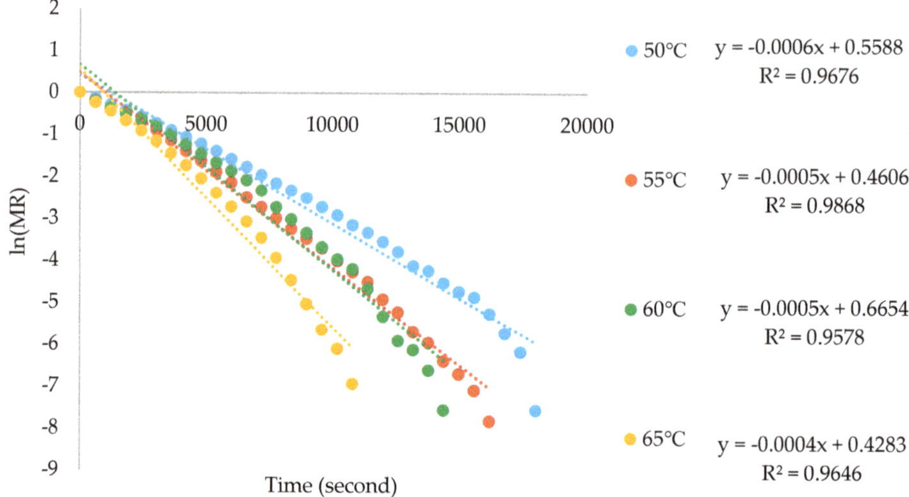

Figure 5. The change of ln(MR) by time (seconds) of sliced lemongrass.

Table 3. D_{eff} at different drying temperatures.

Temperature	Slope	D_{eff} (m/s^2)
50 °C	−0.000353784	7.64×10^{-11}
55 °C	−0.000475707	1.03×10^{-10}
60 °C	−0.000541341	1.17×10^{-10}
65 °C	−0.000684259	1.48×10^{-10}

As shown in Table 3, the D_{eff} value ranged from 7.64089×10^{-11} m^2/s to 1.47784×10^{-10} m^2/s. This result is within the normal value range of D_{eff} in typical food drying processes, which is from 10^{-12} to 10^{-6} m^2/s [14]. At temperatures of 50 °C, 55 °C, 60 °C, and 65 °C, the D_{eff} values were 7.64089×10^{-11}, 1.02741×10^{-10}, 1.16917×10^{-10}, and 1.47784×10^{-10} m^2/s respectively. This indicates that the effective diffusivity coefficient increased proportionally with temperature. At 65 °C, moisture content of citronella peaked, since higher temperature quickens evaporation of water molecules on the surface of the lemongrass material.

3.4. Estimation of Activation Energy

The graph of the change in D_{eff} value (RT_a^{-1}) is displayed in Figure 6. Using exponential regression, the activation energy of E_a of sliced lemongrass was determined to be 38.34 kJ/mol. This result is consistent with the normal range of 33.21 to 39.03 kJ/mol in the drying process of basil leaf [25]. To compare, for activation energy of fruits and vegetables, more than 90% of the activation energy values that were found in previous studies ranged between 14.42 and 43.26 kJ/mol, and 8% of the values were in the range 78.93 to 130.61 kJ/mol [25]. The present result of the activation energy for lemongrass moisture evaporation is relatively high. As a result, the separation of moisture from lemongrass could be difficult, suggesting that, in order to completely dry the lemongrass material, either drying duration or drying temperature should be set at high levels.

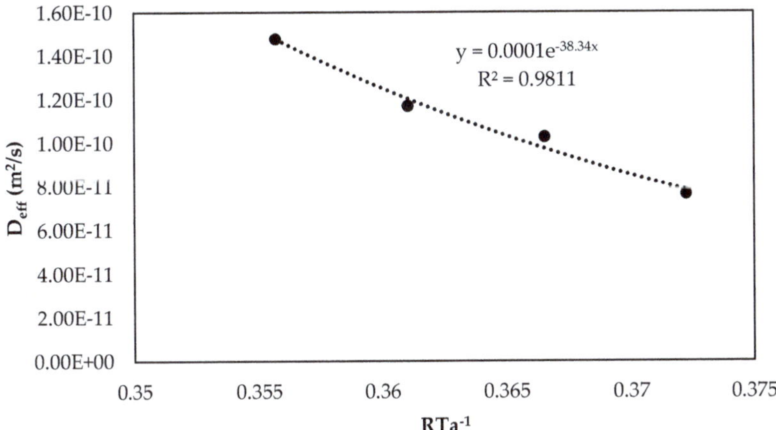

Figure 6. The relationship between change of D_{eff} and $RT_a{}^{-1}$.

4. Conclusions

Our research has identified a mathematical model suitable to describe the behavior of sliced lemongrass subjected to convection drying. In addition, the moisture diffusion coefficient and the activation energy of the material were computed. The results show that higher drying temperature is associated with shorter drying time and faster moisture removal rate. During the drying process, only initial and falling-rate periods occurred, with the absence of a constant-rate period. The Weibull model was found to be the most suitable model describing thin-layer drying by hot air.

The effective moisture diffusivity of the dried material at different temperatures was calculated and the results ranged from 7.64089×10^{-11} m^2/s to 1.47784×10^{-10} m^2/s. The moisture diffusion capability of the lemongrass material peaked at 65 °C. The calculated activation energy of lemongrass was relatively high, at 38.34 kJ/mol, indicating that more energy is needed to separate moisture from the material by drying. This may be related to the structure of the lemongrass, which consists of many layers resulting in tight links among water molecules and in turn low evaporability. These results suggest that moisture removal from the lemongrass should also take into account the remaining content of essential oils and aromatic compounds in the material.

Author Contributions: Investigation, T.V.L.N., D.M.N. and T.D.L.; supervision, L.G.B.; writing—original draft, T.V.L.N.; writing—review & editing, D.C.N.

Funding: This research is funded by Tien Giang Department of Science and Technology, Vietnam.

Acknowledgments: The authors would like to thank Nguyen Tat Thanh University for permission and providing facilities during the research period.

Conflicts of Interest: The authors declare no conflict of interest.

References

1. Skaria, B.P.; Joy, P.P.; Mathew, S.; Mathew, G. Lemongrass. In *Handbook of Herbs and Spices*, 3rd ed.; Peter, K.V., Ed.; Woodhead Publishing Limited: Cambridge, UK, 2006; pp. 400–419.
2. Lindenmuth, G.F.; Lindenmuth, E.B. The Efficacy of Echinacea Compound Herbal Tea Preparation on the Severity and Duration of Upper Respiratory and Flu Symptoms: A Randomized, Double-Blind Placebo-Controlled Study. *J. Altern. Complement. Med.* **2000**, *6*, 327–334. [CrossRef] [PubMed]
3. Haque, A.N.M.A.; Remadevi, R.; Naebe, M. Lemongrass (Cymbopogon): A review on its structure, properties, applications and recent developments. *Cellulose* **2018**, *25*, 5455–5477. [CrossRef]
4. Ekpenyong, C.E.; Akpan, E.; Nyoh, A. Ethnopharmacology, phytochemistry, and biological activities of *Cymbopogon citratus* (DC.) Stapf extracts. *Chin. J. Nat. Med.* **2015**, *13*, 321–337. [CrossRef]

5. Barbosa, L.; Pereira, U.; Martinazzo, A.; Maltha, C.; Teixeira, R.; Melo, E. Evaluation of the Chemical Composition of Brazilian Commercial *Cymbopogon citratus* (D.C.) Stapf Samples. *Molecules* **2008**, *13*, 1864–1874. [CrossRef] [PubMed]

6. Olorunnisola, S.K.; Hammed, A.M.; Simsek, S. Biological properties of lemongrass: An overview. *Int. Food Res. J.* **2014**, *21*, 455–462.

7. Korenblum, E.; Regina de Vasconcelos Goulart, F.; de Almeida Rodrigues, I.; Abreu, F.; Lins, U.; Alves, P.; Blank, A.; Valoni, É.; Sebastián, G.V.; Alviano, D.; et al. Antimicrobial action and anti-corrosion effect against sulfate reducing bacteria by lemongrass (Cymbopogon citratus) essential oil and its major component, the citral. *AMB Express* **2013**, *3*, 44. [CrossRef] [PubMed]

8. Nguyen, H.; Campi, E.M.; Roy Jackson, W.; Patti, A.F. Effect of oxidative deterioration on flavour and aroma components of lemon oil. *Food Chem.* **2009**, *112*, 388–393. [CrossRef]

9. Aguilera, J.M.; Stanley, D.W. *Microstructural Principles of Food Processing and Engineering*, 2nd ed.; Aspen Publishers: Gaithersburg, MD, USA, 1999; ISBN 9780834212565.

10. Doymaz, İ. Drying Kinetics and Rehydration Characteristics of Convective Hot-Air Dried White Button Mushroom Slices. *J. Chem.* **2014**, *2014*, 1–8. [CrossRef]

11. Coradi, P.C.; de Melo, E.C.; da Rocha, R.P. Evaluation of Electrical Conductivity as a Quality Parameter of Lemongrass Leaves (Cymbopogon citratus Stapf) Submitted to Drying Process. *Dry. Technol.* **2014**, *32*, 969–980. [CrossRef]

12. Simha, P.; Mathew, M.; Ganesapillai, M. Empirical modeling of drying kinetics and microwave assisted extraction of bioactive compounds from *Adathoda vasica* and *Cymbopogon citratus*. *Alex. Eng. J.* **2016**, *55*, 141–150. [CrossRef]

13. Association of Official Analytical Chemists. *Official Methods of Analysis*; AOAC: Washington, DC, USA, 1999.

14. Rocha, R.P. Influence of drying process on the quality of medicinal plants: A review. *J. Med. Plants Res.* **2011**, *5*, 7076–7084. [CrossRef]

15. Mohamed Hanaa, A.R.; Sallam, Y.I.; El-Leithy, A.S.; Aly, S.E. Lemongrass (Cymbopogon citratus) essential oil as affected by drying methods. *Ann. Agric. Sci.* **2012**, *57*, 113–116. [CrossRef]

16. Buggle, V.; Ming, L.C.; Furtado, E.L.; Rocha, S.F.R. Influence of Different Drying Temperatures on the Amount of Essential Oils and Citral Content in *Cymbopogon citratus* (DC) Stapf.-Poaceae. *Acta Hortic.* **1999**, *5*, 71–74. [CrossRef]

17. Buchaillot, A.; Caffin, N.; Bhandari, B. Drying of Lemon Myrtle (Backhousia citriodora) Leaves: Retention of Volatiles and Color. *Dry. Technol.* **2009**, *27*, 445–450. [CrossRef]

18. Rocha, R.P.; Ming, L.C.; Marques, M.O.M. Influence of five drying temperatures on the yield and composition of essential oil of citronella (Cymbopogon winterianus Jowitt). *J. Med. Plants* **2000**, *3*, 73–78.

19. Velić, D.; Planinić, M.; Tomas, S.; Bilić, M. Influence of airflow velocity on kinetics of convection apple drying. *J. Food Eng.* **2004**, *64*, 97–102. [CrossRef]

20. Akpinar, E.; Midilli, A.; Bicer, Y. Single layer drying behaviour of potato slices in a convective cyclone dryer and mathematical modeling. *Energy Convers. Manag.* **2003**, *44*, 1689–1705. [CrossRef]

21. Toğrul, H. Suitable drying model for infrared drying of carrot. *J. Food Eng.* **2006**, *77*, 610–619. [CrossRef]

22. Crank, J. *The Mathematics of Diffusion, Oxford Science Publications*, 2nd ed.; Oxford University Press: Oxford, UK, 2009; ISBN 9780198534112.

23. Lopez, A.; Iguaz, A.; Esnoz, A.; Virseda, P. Modelling of sorption isotherms of dried vegetable wastes from wholesale market. *Dry. Technol.* **2000**, *18*, 985–994. [CrossRef]

24. Chong, C.H.; Law, C.L.; Cloke, M.; Hii, C.L.; Abdullah, L.C.; Daud, W.R.W. Drying kinetics and product quality of dried Chempedak. *J. Food Eng.* **2008**, *88*, 522–527. [CrossRef]

25. Onwude, D.I.; Hashim, N.; Janius, R.B.; Nawi, N.M.; Abdan, K. Modeling the Thin-Layer Drying of Fruits and Vegetables: A Review: Thin-layer models of fruits and vegetables. *Compr. Rev. Food Sci. Food Saf.* **2016**, *15*, 599–618. [CrossRef]

26. Ibrahim, M.; Sopian, K.; Daud, W.R.W. Study of the Drying Kinetics of Lemongrass. *Am. J. Appl. Sci.* **2009**, *6*, 1070–1075. [CrossRef]

 *processes*

Article

Digitalizing the Paints and Coatings Development Process

Tomaž Kern *, Eva Krhač, Marjan Senegačnik and Benjamin Urh

Laboratory of Enterprise Engineering, Faculty of Organizational Sciences, University of Maribor, Kidričeva cesta 55a, 4000 Kranj, Slovenia

* Correspondence: tomaz.kern@um.si; Tel.: +386-(0)4-23-74-279

Received: 18 July 2019; Accepted: 12 August 2019; Published: 15 August 2019

Abstract: Numerous laboratory tests are used to determine the appropriateness of new formulations in the development process in the paint and coatings industry. New formulations are most often functionally inadequate, unacceptable for environmental or health reasons, or too expensive. Formulators are obliged to repeat laboratory tests until one of the formulations fulfills the minimum requirements. This is cumbersome, slow, and expensive, and can cause ecological problems, wasting materials on tests that do not produce the desired results. The purpose of this research was to find out if there might be a better way forward to increase efficiency and free up formulators to focus on new products. In this experiment, a new paints and coatings development process was redesigned based on the potential benefits of formulation digitalization. Instead of laboratory testing, a digital platform was used that has been developed and stocked with relevant, up-to-date, and complete, usable data. This study found that, by going digital, developers could vastly reduce non-value-added activities in the development process (by as much as 70%) and significantly shorten the entire process throughput time (by up to 48%). Using digital tools to facilitate the development process appears to be a possible way forward for the paint and coatings industry, saving time, materials, and money and protecting the environment.

Keywords: coatings industry; digitalization; development process; technical enabler; process analysis; process simulation

---

1. Introduction

By analyzing the relevant scientific and technical literature, the most important challenges in the process of developing paints and coatings were identified. Products should meet the aesthetic criteria, as well as assure the long-term protection of the surface. Increasing requirements for effective surface protection, on the one hand, and continuous tightening of environmental regulations, on the other, can be observed [1,2]. An additional focus is achieving new product development and specifications in the shortest possible time. Because of these requirements, rapid development of technology offers more and more opportunities for the design of quality products; however, the development of new products becomes increasingly complex [3–6].

Continually evolving demands for greater capability, efficiency, and functionality of coatings; increasing regulatory complexity; and growing competition in this area account for the need for accelerated development of coatings' raw materials, formulated products, and rapid problem solving [7]. Because of the many raw material combinations and interactions that affect performance properties, additive formulating is resource-intensive and time-consuming for manufacturers [8].

With the requirement for safe, environmentally friendly coating components and the acceleration of material discovery, the need for rapid, reliable testing is becoming more important [9]. A number of accelerated tests have been developed to reduce the testing time and decrease environmental variability. Most importantly, the tests lead to qualitative results, which can be significantly different when various individual evaluators and different test instruments are compared [10].

Coating formulations include different raw materials, such as resins, additives, pigments, fillers, catalysts, solvents, co-solvents, water, etc. Therefore, exploring all possible combinations in search of the best formulation is impossible—especially on the benchtop, which allows for a limited number of options to be explored. To better visualize what that means, the senior manager of R&D Additives (Evonik Resource Efficiency GmbH) presents an example: "If 10 resins, 10 additives, 10 pigments, and 10 catalysts were to be considered for a formulation, 10,000 combinations are possible without considering different quantities for each." It is necessary to explore a large number of possible combinations in order to find the optimal formulation, which is possible only with high-throughput technology [7].

Many papers deal with techniques, such as high-throughput testing [11], statistically planned experiments [12], multiscale modeling, and self-repair, which can make significant contributions to new product development. However, these techniques could be combined in a synergistic manner and thus develop new approaches to self-repair [4]. Data modeling, computer simulation, and visualization tools have also been developed to keep pace with increasingly large datasets. With the aim of the comprehensive evaluation of additives, they suggest the use of high-throughput tools, which facilitate the evaluation of a wider range of samples, usage levels, and formulations [8,13].

Evonik Resource Efficiency GmbH has developed a high-throughput system that automatically doses raw materials, formulates them into coatings, applies the coatings to substrates, and tests them. The system allows the formulation of 120 samples within 24 h. The company emphasizes that the system also completes labor-intensive tasks and thus allows laboratory staff to focus on experimental design and analysis [7]. It is clear that high-throughput regression methods will reduce lead optimization times, but such methods are of little help unless suitable high-quality materials are available for screening, and detailed correlative mechanistic studies are needed to subsequently extract the design principles [14,15]. More important than how much time one repetition of a R&D testing process takes is what to test in order to avoid superfluous testing.

The solution could be the use of information technology and digital transformation approaches. Those approaches are widely and successfully used in other industries [16]. It is time to transfer best practices to the paints and coatings industry, too. The 2019 chief information officer (CIO) of Gartner's agenda [17] recommends that CIOs "secure a new foundation for digital business." The agenda is based on a Gartner survey of 3102 CIO respondents in 89 countries and across major industries, representing $15 trillion in revenue and public sector budgets and $284 billion in information technology (IT) spending. The agenda recommends: (1) a business and operating model transformation, (2) secure consumer-centricity, (3) the use of targeted and capability-building techniques to move from projects to direct product delivery, and (4) the use of business-enabling technologies: machine learning, data analytics (including predictive analytics), cloud processing and storage (including XaaS), digital transformation, and more.

Digitalization is a completely different approach and requires significant changes to take place on many different levels throughout the entire organization and not only for the team responsible for strategy implementation. For the purpose of better serving their customers, organizations must strive to connect all business and management processes [18].

Some early signs that digitalization is becoming fact in the coatings area are already present. However, only a few companies have created digitalization strategies and established business groups focused on digitalization. However, most are adopting this approach only, for example, to improve customer experience or business processes, or to introduce new business models. In 2016, a white paper from the Digital Transformation Initiative for the chemistry and advanced materials industry was published. The paper emphasizes that, with digitalization driving operational optimization, one can expect improved efficiency, productivity, and safety across the whole supply chain of paints and coatings. It is known that digitalization opportunities go beyond and across functional areas, but the question is when they will be enforced. The director of digital and e-commerce says that the answer is dependent on technological maturity, business activities, and the strength and willingness of the company to digitalize, innovate, and change [18].

Some modern tools and technologies already exist, such as "blockchain," "Internet of Things (IoT)," "quantum computing," etc. However, the most important "digital transformation" tools are those that enable data and knowledge management along and across both internal functions and external networks, and that allow companies to find specific pieces of information within the huge ocean of data available in a structured way. Those tools can facilitate smart ways to combine systems that use internal product, customer, market, and manufacturing data. Companies that wish to benefit will also have to make some data available to the system. For example, in the marine coatings area, a "big data" service has become available, which helps select the optimal anti-biofouling coating, based on analysis of billions of data points. At the same time, "cloud-based" platform access and shared external and internal data enable estimated arrival times of ships at port, which helps to optimize delivery planning [15]. Also, a case of prediction profilers using historical data has been published. Prediction profilers can be used in combination with 3D data visualization tools to describe important results. The ability to predict performance reduces the number of formulations that need to be physically tested in a laboratory. Consequently, the development process is accelerated [7].

Both of the above examples are positive. However, a thorough and radical reengineering of key business processes could not be found, especially internal and external processes in a value chain at the same time. A business process is defined as "the sequence of activities in an organizational and technical environment with a structure that describes their logical order and dependence, with the main goal to produce the desired result" [19,20]. New technology can significantly improve effectiveness and efficiency, but it can also make the existing process more complex, reducing usability and causing more integration problems. As processes become more complex, problems with locating and correcting are increasing dramatically [21]. Business process adjustment is often carried out through business process redesign projects, which have the same goal: to achieve more efficient operations. According to Urh, Kern and Roblek [22], top managers are often faced with important questions: What is the level of business process performance efficiency? Is it necessary or reasonable to adapt the process? What adjustments must be made in business process performance? and How will the projected changes influence the business process performance efficiency?

These questions are also asked about the paints and coatings development process. The purpose of this research was to investigate whether the process can be dramatically and rationally improved using cloud-based information technology and a "big data" approach. In order to carry out the research, a sample of several companies involved in the production of paints and coatings was selected. The sample included small and medium-sized enterprises that carry out the development process in a classical manner—without the use of information technology that would enable digital transformation. The following sections describe the basis of the methodologies used (Section 2) and present research results (Section 3). This article concludes with a discussion of the research results and the conclusions reached (Section 4).

2. Methods

Significant progress in the paints and coatings R&D can be made by process re-engineering, not only by improving individual phases (e.g., lab testing). For radical improvement, it is necessary to use business process re-engineering (BPR) or business process management (BPM) and innovative information and communication technologies (ICT), combined with a digitalization approach.

For re-engineering the paints and coatings development process, a process analysis has to be performed for which relevant and up-to-date data has to be obtained, process models developed and a "technological enabler" selected [16]. Finally, the renewed process has to be validated by the simulation.

Process execution data were obtained through targeted interviews and modeling of the development process. In order to obtain the throughput time of individual phases and process activities, a deconstruction was done for each individual activity; into waiting time, orientation time (preparation-finishing time), and processing time [23] (Figure 1). The deconstruction of time was simplified with regard to the structure of activity time, which is stated by Ljubič [24]: processing time, preparation-finishing time, waiting time before and after activity, and transport time to the next activity (Figure 2). Time data can be measured in various time units, for example: seconds, minutes, hours, or days. Hours were used as the time unit in the present research.

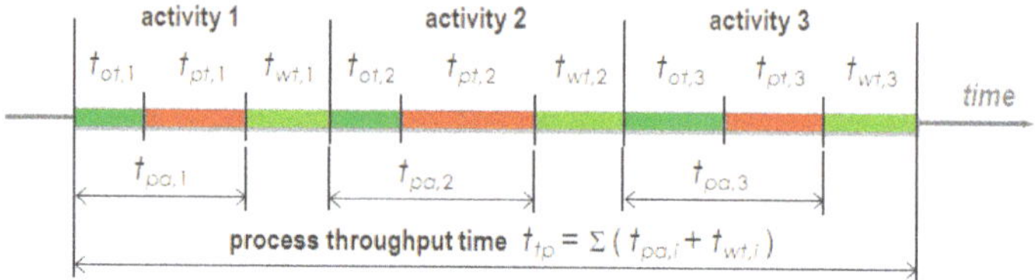

Figure 1. The structure of process throughput time.

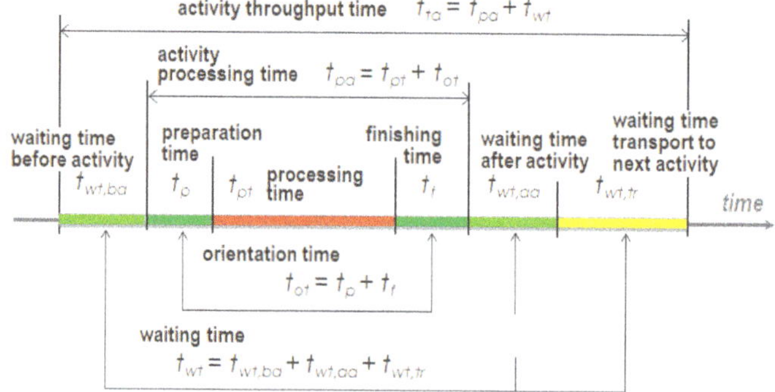

Figure 2. The structure of activity throughput time.

Based on the obtained sample, individual expected times with calculations from the PERT method were predicted [25]: optimistic, most probable, and pessimistic time:

$$t_e = (t_o + 4t_m + t_p)/6 \tag{1}$$

t_e—expected time

t_o—optimistic time

t_m—most probable time

t_p—pessimistic time.

This is a development process with a relatively small number of repetitions, therefore possible errors in determining the actual values of time should not be ignored. However, this error was canceled out, since the same time values were taken into account in the proposal of the renewed process. Thus, the relationship with which the development process was supported with the latest information technology was justified, remains the same.

The business process simulation was conditioned by having process models in the appropriate repository. It is only possible to perform a simulation with a proper set of data. Architecture of integrated information systems (ARIS) methodology, more specifically an event-driven process chain model (EPC model type), was used for modeling because it presents a user perspective of the process [26–28]. This model is based on the logic that an event triggers an activity (task) or several activities. Consequently, the activity ends with a new event or several events.

Standard symbols [23] for business objects and relations were used (an example of a process section is given in Figure 3). The rules for using logical operators are also shown in the example.

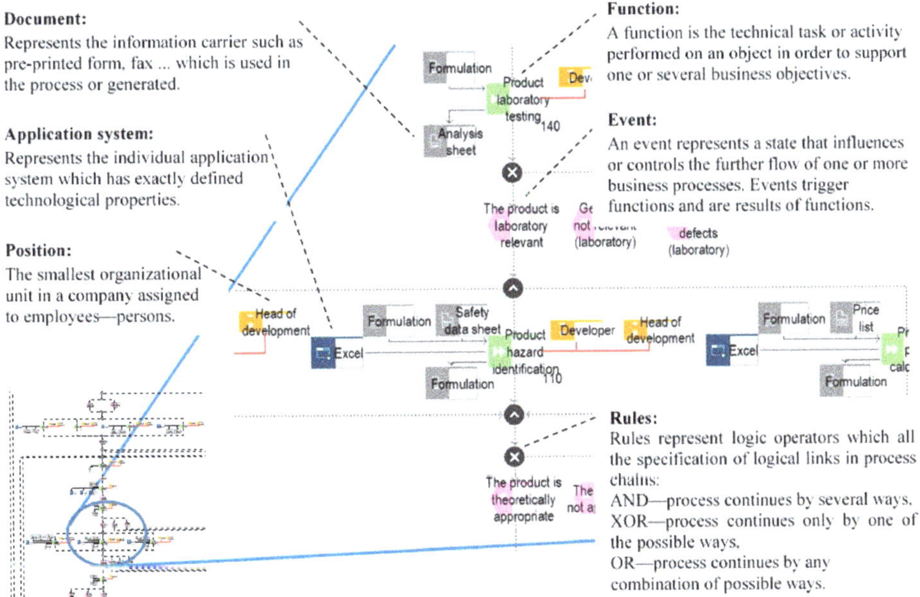

Figure 3. Process example and description of symbols used in business process mapping with the event-driven process chain (EPC) model type.

Through the literature review a lot of recommendations for process performance efficiency assessment were encountered [19,29,30], which can be performed based on the operational and structural efficiency indicators. The first indicators are connected with time, costs, and/or quality [31,32], whereas structural efficiency indicators are connected with business process structural complexity [33–36].

For each performance dimension, it was possible to identify different key performance indicators. Due to the simulation, the focus was on the time dimension and the following indicators were identified [32,37]:

- the waiting time—time spent waiting for a resource to become available. It is possible to measure the waiting time for each activity or for the whole case; attributes specify the wait time accrued for process folders as a rule;
- the service (processing) time—the actual work time put into the case. In the case of concurrency, the total service time (the sum of the time spent on the execution of various activities) may exceed the execution time. However, the service time is usually only part of the execution time; attributes specify the duration of time that may elapse between the start time and end time of a task;
- the orientation time—attributes specify the orientation times required for a function or resource based on the last simulation run;
- the lead (throughput) time—the total time from beginning to end of an individual case of process execution. The average execution time can also be measured, while the level of variance is important: it is not the same if all the cases last for about two weeks or if the individual lasts only a few hours and others for more than one month. The total time for carrying out a function one time should be the sum of the processing time and the orientation time, excluding the wait time.

The simulation is generally an imitation of the operation of a real-world process or system. It is used to assist decision-making by providing a tool that allows the current behavior of a system to be analyzed and understood. It is also able to help predict the performance of that system under a number of scenarios determined by the decision-maker [38]. Simulation modeling has been used for many years in the manufacturing sector and has become a mainstream tool in business situations. This is partly because of the popularity of business process re-engineering (BPR) and other process-based improvement methods that use simulation to help analyze changes in process design [39].

Simulation, in general, covers a large area of interest (e.g., business system performance prediction, providing performance measures). Simulation can refer to a range of model types, from spreadsheet models (static models) to system dynamic and discrete event simulation (dynamic models) [39]:

- static simulation—static models include the linear programming technique, which is an example of an analytical mathematical approach that can be used to solve management decision-making problems. A computer spreadsheet is an example of a numerical static model in which relationships can be constructed and the system behavior studied for different scenarios.
- dynamic simulation—a dynamic mathematical model allows changes in system attributes to be derived as a function of time. The derivation may be made with an analytical solution or with a numerical computation, depending upon the complexity of the model (process). Models that are of a dynamic nature and cannot be solved analytically must use the simulation approach. A classification is made between continuous and discrete event simulation model types. A discrete system changes only at discrete points in time. In practice, most continuous systems can be modeled as discrete at different levels of abstraction.

In the validation and verification of the simulation model, the dynamic simulation was used. Also, the static simulation was used for the impact assessment of the proposed changes in the process implementation. For static simulation execution Excel spreadsheets were used and for dynamic simulation execution the Aris tool was used. Discrete event simulation (DES) in the Aris tool works by modeling individual events that occur using a time-based engine, taking into account resources, constraints, and interaction with other events. This technique can easily reflect the process rules, randomness, and variability that affect the behavior of real-life systems and complex operating environments [40]. The simulation execution steps are as follows:

1. defining simulation objectives;
2. identifying the system/process;
3. collecting and analyzing system/process data;
4. preparation of the simulation model and program;
5. simulation model evaluation;
6. simulation execution;
7. analysis of simulation results;
8. simulation conclusion.

A technical enabler for the proposed improvement of development process has already been developed and is used as an information tool of the fourth generation [41]. The tool is at the stage of prototype testing and, according to this research, is the only all-in-one tool that enables online, real-time searching for raw materials, virtual formulation of paints and coatings, and the creation of digital technical and safety data sheets. It enables the formation of paints and coatings formulations based on data about binders, pigments, additives, and solvents. The formulator uses materials data from the structured database, which is available in digital form in a cloud. The re-engineering point is that the formulator has instant and free access to a large number of raw materials and is guided by the platform to select only those that are functionally relevant, safe, environmentally acceptable, and affordable, even before the individual formulation is laboratory-tested. This method significantly reduces the number of unnecessary laboratory tests and consequently significantly reduces the paints and coatings development throughput time. The advantage of the reengineering is that the data for the product are already generated, available, and ready to use for the preparation of necessary documentation (i.e., safety data sheets, technical data sheets, and hazard labels).

3. Results

Referring to the simulation execution steps, defining simulation objectives (1), identifying the system/process (2), and collecting data (3) have already been defined in previous sections. The research goal was to optimize the throughput time of the paints and coatings development process.

Different development processes (which are performed in practice by individual companies) without the use of the latest cloud-based ICT, can be roughly separated into two variants of the process:

- development of a new product without ICT support (classical process),
- development of a new product with ICT support and local database use.

A dissection of the process execution variants into key activities and the possible support of their execution with appropriate ICT equipment is presented in Table 1.

Table 1. Key activities of process variants and their support with information and communication technologies (ICT).

New Product Development Process without ICT Support (The Classic Process)			New Product Development Process with ICT Support and Using a Local Database		
##	Process Activity	ICT	##	Process Activity	ICT
10	Creating a new product idea		10	Creating a new product idea	
20	Market analysis of existing products		20	Market analysis of existing products	
30	Searching for suitable binders		30	Searching for suitable binders	✓
40	Study of binders' properties		40	Study of binders' properties	✓
50	Searching for pigments		50	Searching for pigments	✓
60	Searching for additives		60	Searching for additives	✓
70	Searching for solvents		70	Searching for solvents	✓
80	Searching for fillers		80	Searching for fillers	✓
90	Formulation (modified) formulations		90	Formulation (modified) formulations	✓
100	Ordering samples		100	Ordering samples	
110	Product laboratory testing		110	Product laboratory testing	
120	Product parameters measurement		120	Product parameters measurement	
130	Product hazard identification		130	Product hazard identification	
140	Product price calculating		140	Product price calculating	
150	Internal validation		150	Internal validation	
160	External validation		160	External validation	
170	Preparation of documentation draft		170	Preparation of documentation draft	✓
180	Creating documentation		180	Creating documentation	

Since the research goal was to analyze the throughput time of the development process execution, data about the structure of individual activity times in the process were collected: waiting time, orientation time, and processing time. In the branching or aggregating processes, the probability assessment was also obtained (Table 2).

On the basis of the collected data and targeted interviews (with leading employees in the processes), the simulation model (4) in the Aris tool was designed. The simulation model is shown in Figure 4. The figure represents a value-added diagram of the existing process in three levels of decomposition. The improvement of the process is presented below.

The process model was validated and verified (5) with a dynamic simulation of the process execution. The process model segment during the simulation and part of the results report are shown in Figure 5.

The simulation was carried out for a period of one working year (231 working days, eight hours per day). The simulation results were completed within 14 min. These theoretical results were compared with the actual results achieved by the companies in question. Also, the transition time for a successfully

Processes **2019**, 7, 539

executed repetition of the process and the number of successful process executions in one year were compared. The simulation results were consistent with the actual data, so the simulation model of the existing state could be verified. A possible error can only be relative to the execution time of individual activities. By collecting data through multiple iterations of an information supported process, this error will be eliminated.

Table 2. Activity time estimates and the probability of new product development process without ICT support.

##	Process Activity	Time Estimates (in Hours, h)	Optimistic	Most Probable	Pessimistic	Expected	Activity Throughput	Probability [1]
10	Creating a new product idea	Waiting	15.00	18.00	52.00	23.17	95.17	0.06
		Orientation	3.00	8.00	45.00	13.33		
		Processing	4400	55.00	88.00	58.67		
20	Market analysis of existing products	Waiting	15.00	18.00	52.00	23.17	95.17	1.00
		Orientation	3.00	8.00	45.00	13.33		
		Processing	44.00	55.00	88.00	58.67		
30	Searching for suitable binders	Waiting	0.20	0.40	1.00	0.47	4.72	0.92
		Orientation	0.10	0.20	0.60	0.25		
		Processing	2.00	4.00	6.00	4.00		
40	Study of binder properties	Waiting	0.20	0.40	1.00	0.47	4.72	1.00
		Orientation	0.10	0.20	0.60	0.25		
		Processing	2.00	4.00	6.00	4.00		
50	Searching for pigments	Waiting	0.20	0.40	1.00	0.47	4.72	
		Orientation	0.10	0.20	0.60	0.25		
		Processing	2.00	4.00	6.00	4.00		
60	Searching for additives	Waiting	0.20	0.40	1.00	0.47	5.72	
		Orientation	0.10	0.20	0.60	0.25		
		Processing	3.00	5.00	7.00	5.00		0.19
70	Searching for solvents	Waiting	0.20	0.40	1.00	0.47	1.80	
		Orientation	0.10	0.20	0.60	0.25		
		Processing	0.50	1.00	2.00	1.08		
80	Searching for fillers	Waiting	0.20	0.40	1.00	0.47	1.80	
		Orientation	0.10	0.20	0.60	0.25		
		Processing	0.50	1.00	2.00	1.08		
90	Formulation (modified) formulations	Waiting	0.00	0.00	0.00	0.00	2.17	1.00
		Orientation	0.00	0.00	0.00	0.00		
		Processing	1.00	2.00	4.00	2.17		
100	Ordering samples	Waiting	0.20	0.40	1.00	0.47	2.88	1.00
		Orientation	0.10	0.20	0.60	0.25		
		Processing	1.00	2.00	4.00	2.17		
110	Product laboratory testing	Waiting	25.00	50.00	172.00	66.17	159.83	0.50
		Orientation	4.00	8.00	16.00	8.67		
		Processing	40.00	80.00	150.00	85.00		
120	Product parameters measurement	Waiting	0.20	0.40	1.00	0.47	1.28	
		Orientation	0.10	0.20	0.60	0.25		
		Processing	0.40	0.50	1.00	0.57		
130	Product hazard identification	Waiting	0.20	0.40	1.00	0.47	6.38	0.40
		Orientation	0.10	0.20	0.60	0.25		
		Processing	2.00	5.00	12.00	5.67		
140	Product price calculating	Waiting	6.10	24.20	48.50	25.23	27.65	
		Orientation	0.10	0.20	0.60	0.25		
		Processing	1.00	2.00	4.00	2.17		
150	Internal validation	Waiting	25.00	50.00	172.00	66.17	291.50	0.50
		Orientation	4.00	8.00	16.00	8.67		
		Processing	100.00	200.00	400.00	216.67		
160	External validation	Waiting	72.00	232.00	644.00	274.00	291.60	0.75
		Orientation	0.60	1.50	3.00	1.60		
		Processing	8.00	16.00	24.00	16.00		
170	Preparation of documentation draft	Waiting	1.10	2.50	5.00	2.68	8.05	1.00
		Orientation	0.20	1.00	2.00	1.03		
		Processing	2.00	4.00	8.00	4.33		
180	Creating documentation	Waiting	4.10	8.50	17.00	9.18	27.88	1.00
		Orientation	0.20	1.00	2.00	1.03		
		Processing	10.00	16.00	32.00	17.67		

[1] The probability of executing an individual activity according to the decisions during the process. Except in the case of activity "10," there was a probability that the activity was included in the process. In merged fields, probabilities for activities that were executed in parallel (at the same time) were found.

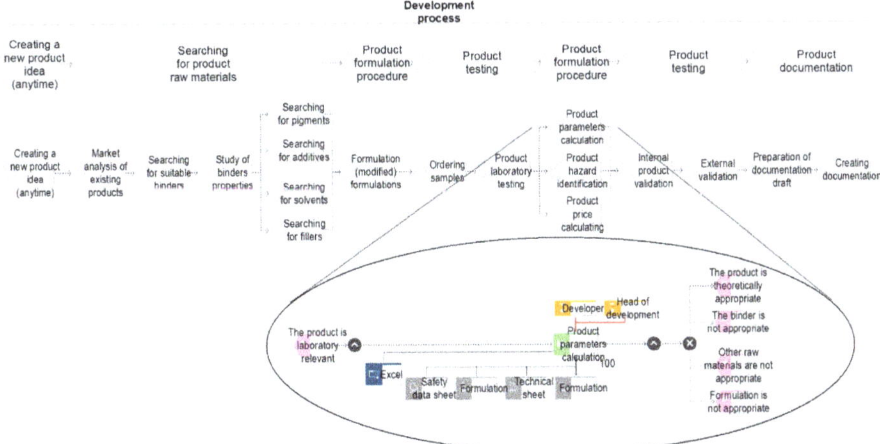

Figure 4. Model overview of the existing new product development process without the support of ICT. (Due to the size of the process model, only two segments of the detailed process model are shown; the entire model is shown in Appendix A).

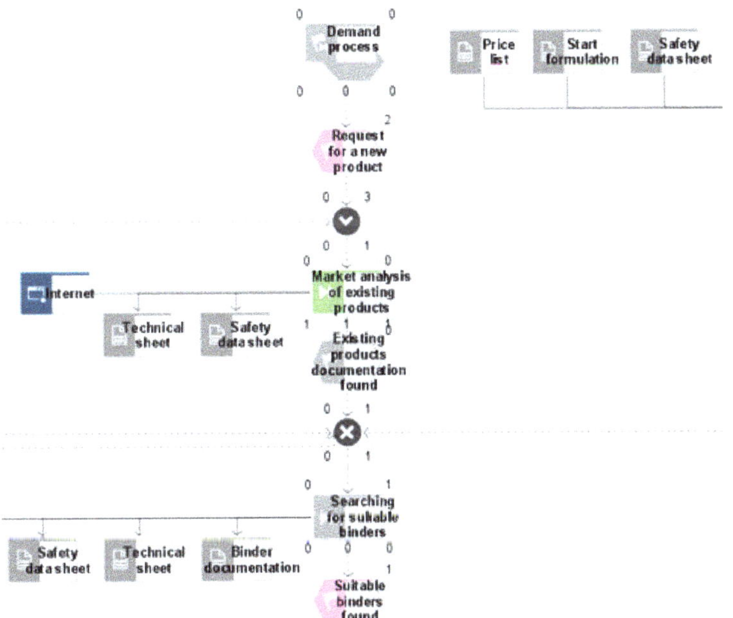

Process instance	Start	End	Throughput time	Static wait time sum	Orientation time sum	Processing time sum	Wait time sum
1	0:00:00	1442:17:19	1442:17:19	818:06:30	35:42:28	597:49:34	248:52:14
2	525:36:00	1587:15:59	1061:39:59	584:57:08	30:32:30	450:34:37	116:49:07
3	0:00:00	1666:45:30	1666:45:30	795:41:40	88:14:59	791:40:09	254:21:25
4	0:00:00	1804:32:21	1804:32:21	874:04:56	72:14:49	864:37:09	158:51:51
5	0:00:00			866:33:07	106:07:34	864:19:44	605:32:06

Figure 5. The segment of the existing process model and part of the dynamic simulation results.

For the purpose of simulating (6) the impact of the proposed changes on the process execution, an appropriate model of the renewed process was developed (shown in Figure 6). It took into account the possibilities offered by the inclusion of process support with modern ICT and the use of a cloud-based database, which allowed for the changed sequence of key activities execution in the process (Table 3). This significantly influenced the time required for the successful execution of the renewed process.

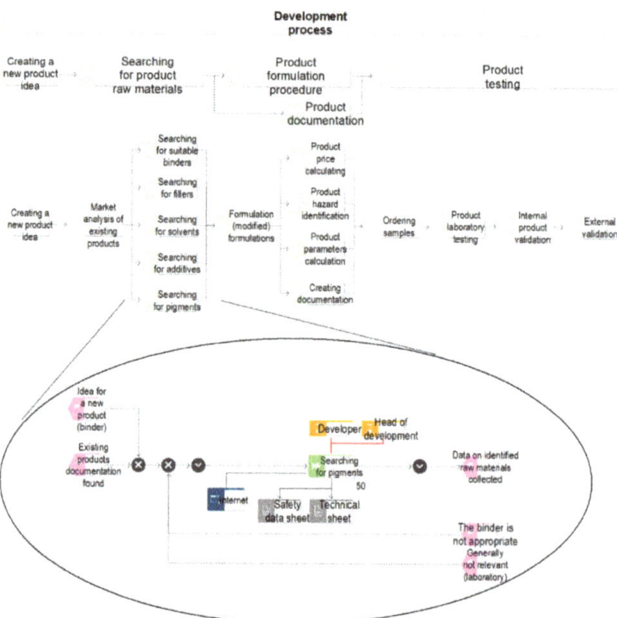

Figure 6. The model overview of the modified (renewed) new product development process with ICT support and the use of a cloud-based database. (Due to the size of the process model, only one segment of the detailed process model is shown; the entire model is shown in Appendix B).

Table 3. Key activities of the new product development process with ICT support and the use of cloud-based database.

New Product Development Process with ICT Support and the Use of a Cloud-Based Database		
##	**Process Activity**	**ICT**
10	Creating a new product idea	✓
20	Market analysis of existing products	
30	Searching for suitable binders	✓
40	Searching for pigments	✓
50	Searching for additives	✓
60	Searching for solvents	✓
70	Searching for fillers	✓
80	Formulation (modified) formulations	✓
90	Product parameters calculation	✓
100	Product hazard identification	✓
110	Product price calculating	✓
120	Creating documentation	✓
130	Ordering samples	✓
140	Product laboratory testing	
150	Internal validation	
160	External validation	

The process was checked in a medium-sized company which develops and produces paints and coatings using this technical enabler [41]. The tool is used to help with the safety data sheets preparation in developing new products [42]. The company can eliminate formulations that will be inadequate, in terms of environmental hazards or cost, in the first stage of the development process. It must be noted that the local database includes a finite amount of the raw materials data that are currently available. The results of the process verification confirmed the simulation results.

The impact of changes on the throughput time was verified by static simulation execution. The simulation of a new product development process without ICT support (taking into account the data in Table 2), as well as a simulation of the renewed process with cloud-based ICT support (taking into account the data in Table 4), were executed.

Table 4. Activity time estimates and the probability of the new product development process with ICT support and the use of a cloud-based database.

##	Process Activity	Time Estimates (in Hours, h)	Optimistic	Most Probable	Pessimistic	Expected	Activity Throughput	Probability 1
10	Creating a new product idea	Waiting	13.50	16.20	46.80	20.85	85.65	0.06
		Orientation	2.70	7.20	40.50	12.00		
		Processing	39.60	49.50	79.20	52.80		
20	Market analysis of existing products	Waiting	15.00	18.00	52.00	23.17	95.17	0.92
		Orientation	3.00	8.00	45.00	13.33		
		Processing	44.00	55.00	88.00	58.67		
30	Searching for suitable binders	Waiting	0.02	0.04	0.10	0.05	0.47	
		Orientation	0.01	0.02	0.06	0.03		
		Processing	0.20	0.40	0.60	0.40		
40	Searching for pigments	Waiting	0.02	0.04	0.10	0.05	0.47	
		Orientation	0.01	0.02	0.06	0.03		
		Processing	0.20	0.40	0.60	0.40		
50	Searching for additives	Waiting	0.02	0.04	0.10	0.05	0.57	0.19
		Orientation	0.01	0.02	0.06	0.03		
		Processing	0.30	0.50	0.70	0.50		
60	Searching for solvents	Waiting	0.02	0.04	0.10	0.05	0.18	
		Orientation	0.01	0.02	0.06	0.03		
		Processing	0.05	0.10	0.20	0.11		
70	Searching for fillers	Waiting	0.02	0.04	0.10	0.05	0.18	
		Orientation	0.01	0.02	0.06	0.03		
		Processing	0.05	0.10	0.20	0.11		
80	Formulation (modified) formulations	Waiting	0.00	0.00	0.00	0.00	0.22	1.00
		Orientation	0.00	0.00	0.00	0.00		
		Processing	0.10	0.20	0.40	0.22		
90	Product parameters calculation	Waiting	0.00	0.00	0.00	0.00	0.00 2	
		Orientation	0.00	0.00	0.00	0.00		
		Processing	0.00	0.00	0.00	0.00		
100	Product hazard identification	Waiting	0.00	0.00	0.00	0.00	0.00 2	0.80
		Orientation	0.00	0.00	0.00	0.00		
		Processing	0.00	0.00	0.00	0.00		
110	Product price calculating	Waiting	1.22	4.84	9.70	5.05	5.53	
		Orientation	0.02	0.04	0.12	0.05		
		Processing	0.20	0.40	0.80	0.43		
120	Creating documentation	Waiting	2.05	4.25	8.50	4.59	13.94	
		Orientation	0.10	0.50	1.00	0.52		
		Processing	5.00	8.00	16.00	8.83		
130	Ordering samples	Waiting	0.04	0.08	0.20	0.09	0.58	1.00
		Orientation	0.02	0.04	0.12	0.05		
		Processing	0.20	0.40	0.80	0.43		
140	Product laboratory testing	Waiting	18.75	37.50	129.00	49.63	119.88	0.50
		Orientation	3.00	6.00	12.00	6.50		
		Processing	30.00	60.00	112.50	63.75		
150	Internal validation	Waiting	25.00	50.00	172.00	66.17	291.50	0.50
		Orientation	4.00	8.00	16.00	8.67		
		Processing	100.00	200.00	400.00	216.67		
160	External validation	Waiting	72.00	232.00	644.00	274.00	291.60	0.75
		Orientation	0.60	1.50	3.00	1.60		
		Processing	8.00	16.00	24.00	16.00		

[1] The probability of executing an individual activity according to the decisions during the process. Except in the case of activity "10," there was a probability that the activity was included in the process. In merged fields, probabilities for activities that were executed in parallel (at the same time) were found. [2] On the basis of an automatic calculation between the paint and coating formulations, the activity does not need extra time, so the parameter value is 0.

Table 5 presents the results of the executed simulation (7) for the new product development process without ICT support and for the renewed process with appropriate ICT support and a cloud-based database.

Table 5. Results of the new product development process static simulation without ICT support and results with ICT support and the use of a cloud-based database.

New Product Development Process without ICT Support (The Classic Process)					
##	Process Activity	Activity Throughput Time	Throughput Time for One Product	With ICT Impact on Execution Time	Without ICT Impact on Execution Time
10	Creating a new product idea	95.17	13.63	13.63	
20	Market analysis of existing products	95.17	218.09		218.09
30	Searching for suitable binders	4.72	10.81	10.81	
40	Study of binder properties	4.72	11.79	11.79	
50	Searching for pigments				
60	Searching for additives				
70	Searching for solvents	5.72	14.29	14.29	
80	Searching for fillers				
90	Formulation (modified) formulations	2.17	28.89	28.89	
100	Ordering samples	2.88	38.44	38.44	
110	Product laboratory testing	159.83	2131.11	2131.11	
120	Product parameters measurement				
130	Product hazard identification	27.65	184.33	184.33	
140	Product price calculating				
150	Internal validation	291.5	777.33		777.33
160	External validation	291.6	388.8		388.8
170	Preparation of documentation draft	8.05	8.05	8.05	
180	Creating documentation	27.88	27.88	27.88	
TOTAL FOR ONE SUCCESSFUL PRODUCT:			**3853.46**	**2469.23**	**1384.22**
##	Process Activity	Activity Throughput Time	Throughput Time for One Product	With ICT Impact on Execution Time	Without ICT Impact on Execution Time
10	Creating a new product idea	85.65	6.13	6.13	
20	Market analysis of existing products	95.17	109.05		109.05
30	Searching for suitable binders				
40	Searching for pigments				
50	Searching for additives	0.57	0.71	0.71	
60	Searching for solvents				
70	Searching for fillers				
80	Formulation (modified) formulations	0.22	1.44	1.44	
90	Product parameters calculation				
100	Product hazard identification				
110	Product price calculating	13.94	92.94	92.94	
120	Creating documentation				
130	Ordering samples	0.58	3.08	3.08	
140	Product laboratory testing	119.88	639.33	639.33	
150	Internal validation	291.5	777.33		777.33
160	External validation	291.6	388.8		388.8
TOTAL FOR ONE SUCCESSFUL PRODUCT:			**2018.82**	**743.65**	**1275.18**
THROUGHPUT TIME REDUCTION (in %):			**47.61**	**69.88**	**7.88**

Based on the comparison of both simulations results (Table 5), the following conclusions about the executed simulation can be drawn (8); the inclusion of the appropriate ICT support and cloud-based database in the new product development process execution results in the following:

- the entire development process throughput time was reduced by 47.61% (1835 out of 3853 h) as is shown in Figure 7,
- the activity throughput time on which the cloud-based ICT has impact was reduced by 69.88% (1726 out of 2469 h),
- surprisingly, even the activity throughput time on which the cloud-based ICT has no impact, due to relevant information from the previous activities, was reduced by 7.88% (109 out of 1384 h).

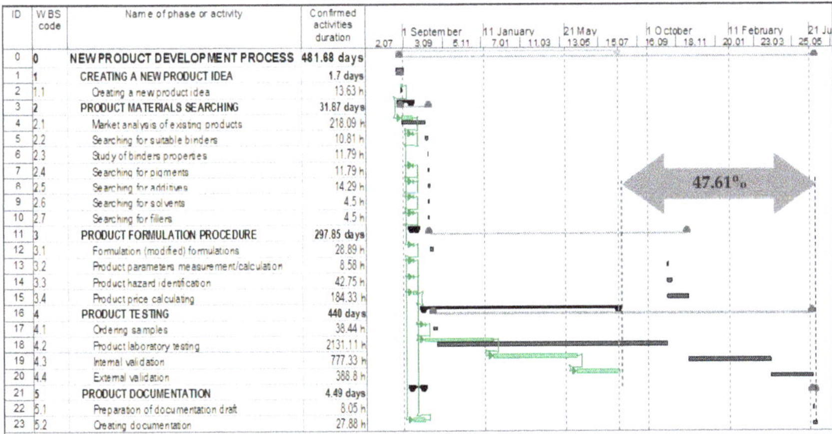

Figure 7. Gantt chart comparison of the existing new product development process (gray) and the modified (renewed) process (green).

Additionally, some other advantages of the renewed process were observed: environmental friendliness, lower cost, innovative, broader offerings, the possibility of manufacturing excellent (optimal) products, the ability to track progress, and a greater possibility of manufacturing niche products in smaller series.

A reduced number of laboratory tests leads to efficient reduction of the waste quantity generated during laboratory tests. Therefore, it helps to reduce the pollution degree (environmental advantage). Laboratory tests are much more expensive than computer simulations. The reason is the price of equipment, human work, energy, and materials (cost advantage). The repetitive work performed by formulators in the laboratory is tiring. With a reduced number of repetitions, formulators have more time to develop new products (innovative advantage).

Smaller companies that have insufficient laboratory capacities can also develop excellent products; there are 3865 such companies in the EU [43] (broader offerings). When the raw materials database is complete (containing thousands or tens of thousands of raw materials), the possibility that optimal raw materials will be selected for a certain product is much greater than in cases in which a formulator executes laboratory tests only from a selection of dozens or hundreds of raw materials (e.g., only those with which they are familiar or those that exist in their local database (manufacturing optimal products)). When the data exist in a global database, manufacturers have their own interest in constantly updating that database. In such a way, companies always have access to the most up-to-date raw materials data. The data changes are accessible to all users (the ability to track progress). The demand for such products increases constantly. In such a manner, it is possible to work on the engineering-to-order (EtO) principle and to use the mass customization principle (manufacturing niche products).

4. Discussion and Conclusions

In this research (which was performed in 2018), models of paints and coatings development processes in international companies were mapped. Data about development process's throughput times and data about the structure of process activities throughput times were collected. Moreover, data about return-loops in processes were ascertained. Significant differences were found between ICT-supported processes and processes executed in the classical way.

Static and dynamic temporal analyses were prepared. The analysis results showed that the most time was saved by a reduction in often-repetitious and time- and resource-consuming laboratory testing activities. According to customer requirements for products with better functionality, laboratory testing activities are necessary. Additionally, there is an increasing supply of raw materials (binders, pigments, additives, solvents and fillers), which lead to more complex development processes. Environmental acceptability and product price are determined only after the laboratory tests (in the classic process), and because of that, more laboratory tests in the development process are necessary. If, upon the completion of a laboratory test, all these properties are not within the limits of expectation and product functionality (both in terms of its functional use and its potential success on the market), the development process returns to the beginning. With requests for shorter processes, companies are faced with a difficult problem: "How can customer requests be fulfilled in a more streamlined manner?"

Companies try to solve the problem by the shortening of individual laboratory tests. In the literature, one can find many innovative solutions that companies either already use or plan to introduce. The majority of these ideas are connected with new technologies or equipment. This can enable the shortening of individual laboratory tests and consequently shorten the development process. However, this is not the final solution to the problem. Because of the increasing complexity of development processes, the implementation of technological changes is not sufficient. Too many, ultimately unnecessary, laboratory tests are regularly undertaken, which slows down the process and can be wasteful in terms of time and materials.

In the present study, an alternative approach to problem resolution was explored. This was the digitalization concept, which is already used in other industries and has yielded excellent results. The fact that technology itself cannot offer the final solution to the problem was also taken into consideration. Technology can only become the "technical enabler," which leads to a renovation of the processes, allowing them to become radically more efficient. To this end, two knowledge areas were explored in detail: new technologies and business processes renovation.

The latest findings on business processes renovation were reviewed. It was found that radical improvements can be achieved by reducing the number of activities. It was also ascertained that laboratory tests are often repeated, and they are on the processes critical path. Accordingly, the focus was shifted to the reduction of the number of laboratory tests, which can be accomplished through adequate selection of formulations ahead of time, before entering the lab. As the environmental acceptability and product price, mainly depend on the raw materials, these can be calculated without laboratory testing. Therefore, it is possible to find which formulations are not acceptable from an environmental or economic standpoint, and they can be excluded from further consideration. Only those formulations that are environmentally and economically acceptable can be included in laboratory testing. This leads to a radical reduction in laboratory testing repetitions and also enables efficient reduction of the development process throughput time.

However, a necessary condition for the approach mentioned above is digital access to relevant, up-to-date, and complete, useable data on raw materials. It was explored whether accessible databases exist and it was found that the majority of data from various manufacturers are publicly available only in the form of safety data sheets, which must be enclosed with the substances or products. However, these data are mainly available in unstructured forms (PDF format), and as such are useless for formulation purposes, because combining data in unstructured documents is time-consuming and impractical. Formulators prefer to execute laboratory tests because that is often easier than searching through a large quantity of data on raw materials, which then will not be laboratory-relevant.

Therefore, technology that enabled rapid, efficient access to relevant data was explored and an ICT solution that supports the development process through the creation of technical documentation (safety data sheets and print labels) found. This technical enabler is unique because it not only supports the creation of technical documentation in PDF format (based on local documents), but also enables the use of global databases. These global databases could be filled with data from raw materials manufacturers and are used by companies that produce paints and coatings when they prepare safety data sheets. Safety data sheets are not in PDF format, but exist in a structured form. Technical documentation is thus only presented as a printout. Formulators can prepare formulations and simultaneously observe all product characteristics. When data about raw material prices are available, the product price can also be calculated.

It was found that the technical enabler could be used in the initial stages of the development process, when the formulator designs the formulation. Instead of executing laboratory tests on a large quantity of formulations and checking the environmental and economic acceptability of formulations only after laboratory tests have been completed, the process was reversed. In the beginning, simulations of possible formulations were prepared using the technical enabler platform. Then, only those formulations that were within the range of tolerance, in terms of hazards and price points, were included in analog laboratory testing. The result was a large number of virtual simulations and a small number of actual laboratory tests. When the data for simulations were accessible, up-to-date simulations could be performed in a very short time and with no waste in terms of materials and non-value-added time. Therefore, the "vision" is to encourage raw materials manufacturers to import the product data into the cloud-based database, where data will be accessible to all formulators. When critical mass is achieved, the greater advantages of this approach will become evident in practice.

Author Contributions: Conceptualization, T.K.; methodology, T.K. and B.U.; software, B.U.; validation, T.K., E.K., and B.U.; formal analysis, E.K. and B.U.; investigation, E.K., M.S., and B.U.; resources, T.K.; data curation, T.K. and B.U.; writing—original draft preparation, E.K., M.S., and B.U.; writing—review and editing, T.K.; visualization, B.U.; supervision, T.K.; project administration, T.K. All authors (T.K., E.K., M.S., and B.U.) read and approved the final manuscript.

Funding: This research received no external funding.

Acknowledgments: The authors are grateful to the Laboratory of Enterprise Engineering, Faculty of Organizational Sciences, University of Maribor for supporting the project. The authors would also like to thank Noah Charney for proof-reading the article.

Conflicts of Interest: The authors declare no conflict of interest.

Appendix A

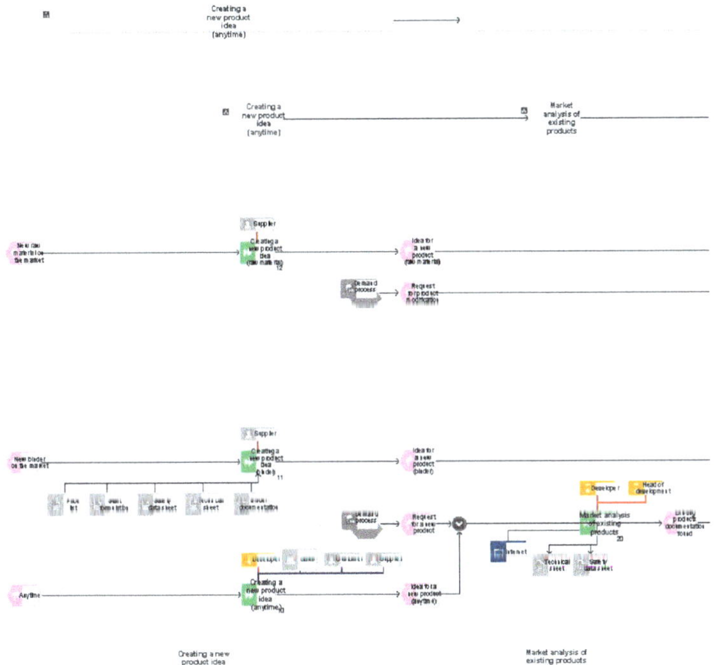

Figure A1. Model overview of the existing new product development process without the support of ICT (part 1).

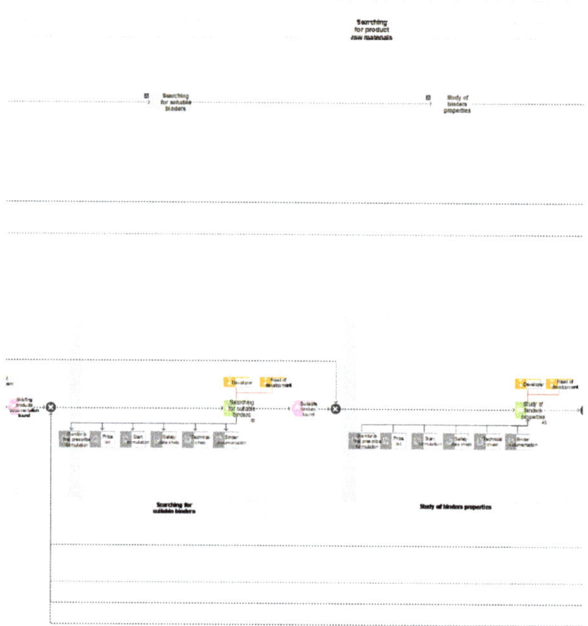

Figure A2. Model overview of the existing new product development process without the support of ICT (part 2).

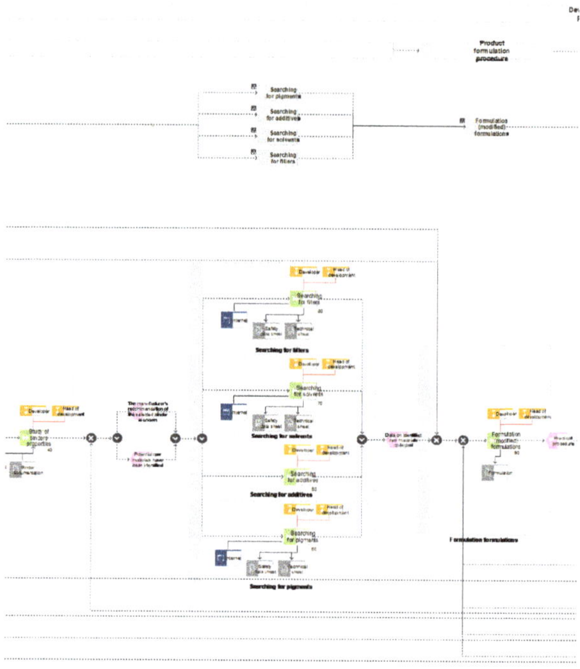

Figure A3. Model overview of the existing new product development process without the support of ICT (part 3).

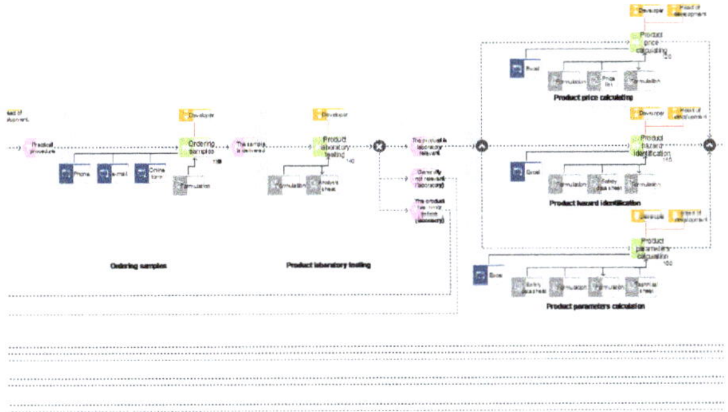

Figure A4. Model overview of the existing new product development process without the support of ICT (part 4).

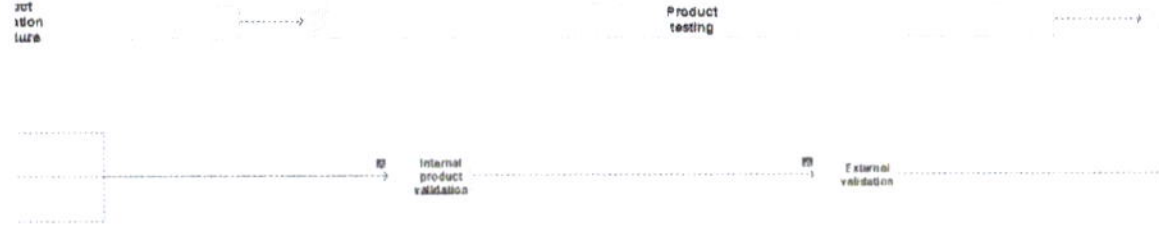

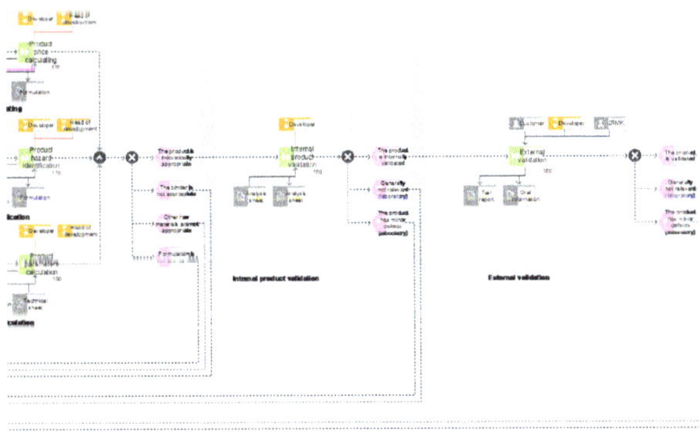

Figure A5. Model overview of the existing new product development process without the support of ICT (part 5).

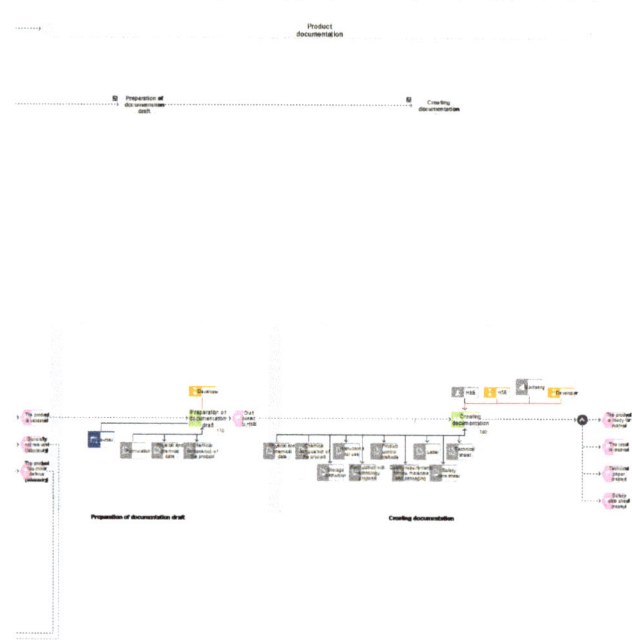

Figure A6. Model overview of the existing new product development process without the support of ICT (part 6).

Appendix B

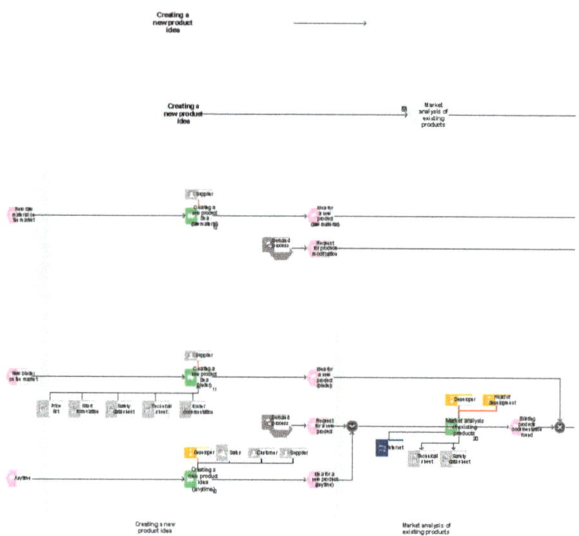

Figure A7. The model overview of the modified (renewed) new product development process with ICT support and the use of a cloud-based database (part 1).

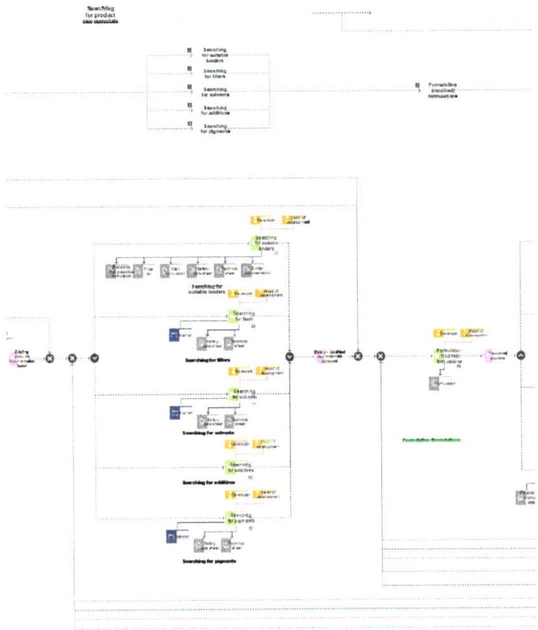

Figure A8. The model overview of the modified (renewed) new product development process with ICT support and the use of a cloud-based database (part 2).

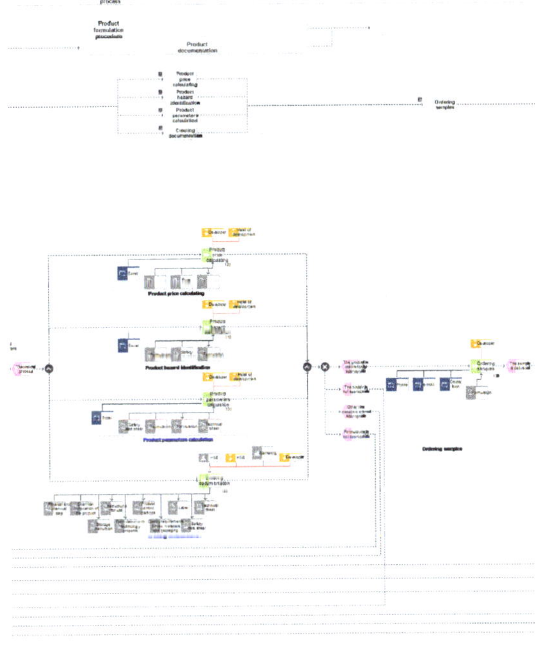

Figure A9. The model overview of the modified (renewed) new product development process with ICT support and the use of a cloud-based database (part 3).

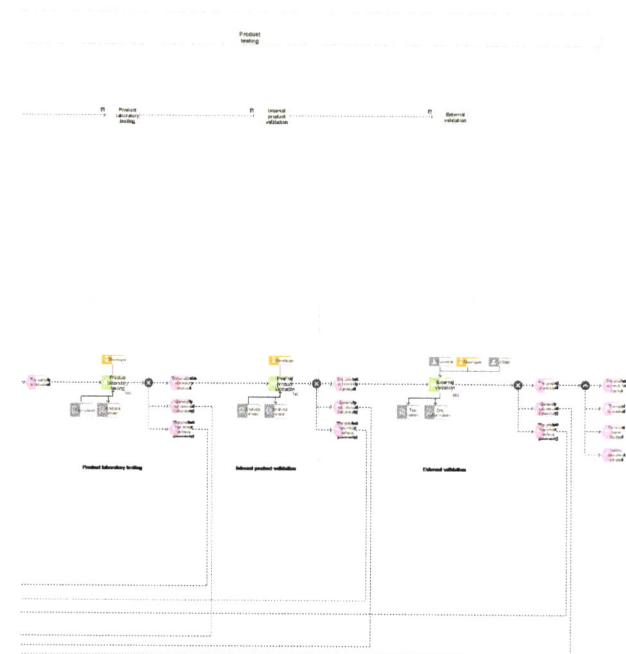

Figure A10. The model overview of the modified (renewed) new product development process with ICT support and the use of a cloud-based database (part 4).

References

1. Nurioglu, A.G.; Esteves, A.C.C. Non-toxic, non-biocide-release antifouling coatings based on molecular structure design for marine applications. *J. Mater. Chem. B* **2015**, *3*, 6547–6570. [CrossRef]
2. Akafuah, N.K.; Poozesh, S.; Salaimeh, A.; Patrick, G.; Lawler, K.; Saito, K. Evolution of the automotive body coating process—A review. *Coatings* **2016**, *6*, 24. [CrossRef]
3. Staring, E.; Dias, A.A.; Van Benthem, R.A. New challenges for R&D in coating resins. *Prog. Org. Coat.* **2002**, *45*, 101–117. [CrossRef]
4. Cole, I.S.; Hughes, A.E. Designing molecular protection: New paradigm for developing corrosion resistant materials uniting hightroughput studies, multiscale modelling and self-repair. *Corros. Eng. Sci. Technol.* **2014**, *49*, 109–115. [CrossRef]
5. Bohorquez, S.J.; van den Berg, P.; Akkerman, J.; Mestach, D.; van Loon, S.; Repp, J. High-throughput paint optimisation by use of a pigment-dispersing polymer. *Surf. Coat. Int.* **2015**, *98*, 85–89.
6. Björklund, S.; Goel, S.; Joshi, S. Function-dependent coating architectures by hybrid powder suspension plasma spaying: Injectordesign, processing and concept validation. *Mater. Des.* **2018**, *142*, 56–65. [CrossRef]
7. Coatingstech, Coatings Xperience: Accelerating Coatings Development with High Troughput Technology. Available online: https://www.paint.org/article/accelerating-coatings-development-high-throughput-technology/ (accessed on 29 October 2018).
8. Langille, M.; Izmitli, A.; Lan, T.; Agrawall, A.; Liu, C.; Henderson, K.; Lu, Y.J. Balancing performance of slip/mar additives using a high throughput approach. *Coatingstech* **2018**, *14*, 52–56.
9. Nameer, S.; Johansson, M. Fully bio-based aliphatic thermoset polyesters via self-catalyzed self-condensation of multifunctional epoxy monomers directly extracted from natural sources. *J. Coat. Technol. Res.* **2017**, *14*, 757–765. [CrossRef]
10. Taylor, S.R.; Contu, F.; Calle, L.M.; Curran, J.P.; Li, W. Predicting the long-term field performance of coating systems on steel using a rapid electrochemical test: The damage tolerance test. *Corros. J. Sci. Eng.* **2012**, *68*. [CrossRef]

11. Wu, L.; Baghdachi, J. *Functional Polymer Coatings: Principles, Methods, and Applications*, 1st ed.; John Wiley & Sons: Hoboken, NJ, USA, 2015.

12. Qureshi, S.A.; Shafeeq, A.; Ijaz, A.; Butt, M.M. Development and regression modeling dirt resistive latex façade paint. *Coatings* **2019**, *9*, 150. [CrossRef]

13. Javierre, E. Modeling self-healing mechanisms in coatings: Approaches and perspectives. *Coatings* **2019**, *9*, 122. [CrossRef]

14. Dennis, R.V.; Patil, V.; Andrews, J.L.; Aldinger, J.P.; Yaday, G.D.; Banarjee, S. Hybrid nanostructured coatings for corrosion protection of base materials: A sustainability perspective. *Mater. Res. Express* **2015**, *2*. [CrossRef]

15. Prat, D.; Wells, A.; Hayler, J.; Sneddon, H.; McElroy, R.; Abou-Shehada, S.; Dunn, P.J. CHEM21 selection guide of classical- and less-classical-solvents. *Green Chem.* **2016**, *18*. [CrossRef]

16. DKE Deutsche Kommission Elektrotechnik Elektronik Informationstechnik in DIN und VDE, German Standardization Roadmap, Industrie 4.0. Available online: https://www.din.de/blob/65354/57218767bd6da1927b181b9f2a0d5b39/roadmap-i4-0-e-data.pdf (accessed on 30 November 2018).

17. Gartner, the 2019 CIO Agenda: Securing a New Foundation for Digital Business. Available online: https://www.gartner.com/doc/3891665/cio-agenda-securing-new-foundation (accessed on 30 November 2018).

18. Challener, C. The paint and coatings industry in the age of digitalization. *Jct Coat.* **2018**, *15*, 54–60.

19. Aguilar-Savén, R.S. Business process modelling: Review and framework. *Int. J. Prod. Econ.* **2004**, *90*, 129–149. [CrossRef]

20. Weske, M. *Business Process Management: Concepts, Languages, Architectures*, 1st ed.; Springer: Berlin, Germany, 2007.

21. Cheng, C. Complexity and Usability Models for Business Process Analysis. Doctoral Dissertation, The Graduate School College of Engineering, Pennsylvania State University, PA, USA, 21 August 2008.

22. Urh, B.; Kern, T.; Roblek, M. *Business Process Modification Management, In Encyclopedia of Networked and Virtual Organizations*, 1st ed.; Putnik, G., Cunha, M.M., Eds.; Information Science Reference: Hershey, CA, USA; New York, NY, USA, 2008; pp. 112–120.

23. Davis, R. *ARIS Design Platform: Advanced Process Modelling and Administration*, 1st ed.; Springer: Berlin, Germany, 2008.

24. Ljubič, T. *Operational Management of Production*, 1st ed.; Modern organization: Kranj, Slovenia, 2006.

25. Cottrell, W.D. Simplified Program Evaluation and Review Technique (PERT). *J. Constr. Eng. Manag.* **1999**, *125*, 16–22. [CrossRef]

26. Scheer, A.W. *ARIS—Business Process Framework*, 2nd ed.; Springer: Berlin, Germany, 1998.

27. Pavlović, I.; Kern, T.; Miklavčič, D. Comparison of paper-based and electronic data collection process in clinical trials: Costs simulation study. *Contemp. Clin. Trials* **2009**, *30*, 300–316. [CrossRef]

28. Sánchez González, L.; García Rubio, F.; Ruiz González, F.; Piattini Velthuis, M. Measurement in business processes: A systematic review. *Bus. Process Manag. J.* **2010**, *16*, 114–134. [CrossRef]

29. Frederiksen, H.; Mathiassen, L. A Contextual Approach to Improving Software Metrics Practices. *IEEE Trans. Eng. Manag.* **2008**, *55*, 602–616. [CrossRef]

30. Sharma, A. Implementing Balance Scorecard for Performance Measurement. *ICFAI J. Bus. Strategy* **2009**, *6*, 7–16.

31. Valiris, G.; Glykas, M. Business analysis metrics for business process redesign. *Bus. Process Manag. J.* **2004**, *10*, 445–480. [CrossRef]

32. van der Aalst, W.M.P. Business process management: A comprehensive survey. *ISRN Softw. Eng.* **2013**, *2013*. [CrossRef]

33. Rolón, E.; Ruiz, F.; Garcia, F.; Piatiini, M. Applying Software Metrics to evaluate Business Process Models. *CLEI Electron. J.* **2006**, *9*, 1–15. [CrossRef]

34. Mendling, J.; Moser, M.; Neumann, G.; Verbeek, H.M.W.; Van Dongen, B.F.; van der Aalst, W.M.P. A Quantitative Analysis of Faulty EPCs in the SAP Reference Model. *BPM Rep.* **2006**, *0608*, 1–49.

35. Cardoso, J.; Mendling, J.; Neumann, G.; Reijers, H.A. A discourse on complexity of process models. In *Business Process Management Workshop*, 1st ed.; Eder, J., Dustdar, S., Eds.; Springer: Berlin, Germany, 2006; pp. 117–128.

36. Cardoso, J. Business Process Control-Flow Complexity: Metric, Evaluation and Validation. *Int. J. Web Serv. Res.* **2008**, *5*, 49–76. [CrossRef]

37. Determine Your Process Costs and Times (Klein, E.). Available online: https://www.ariscommunity.com/users/eva-klein/2011-01-17-determine-your-process-costs-and-times (accessed on 4 December 2018).

38. Banks, J.; Carson, J.S.; Nelson, B.L.; Nicol, D.M. *Discrete-Event System Simulation*, 5th ed.; Pearson Prentice Hall: Upper Saddle River, NJ, USA, 2010.

39. Greasley, A. The case for the organisational use of simulation. *J. Manuf. Technol. Manag.* **2004**, *15*, 560–566. [CrossRef]

40. Lanner Group Ltd, What Is Predictive Simulation? Available online: https://www.lanner.com/en-gb/pages/what-is-predictive-simulation.html (accessed on 25 October 2018).

41. Allchemist. Available online: https://allchemist2018.av-studio.agency/ (accessed on 16 January 2019).

42. ECHA, European Chemicals Agency. Available online: https://echa.europa.eu/regulations/reach/legislation (accessed on 25 January 2019).

43. Eurostat, Industry by Employment Size Class—Economical Indicator for Structural Business Statistics—Enterprises Number. Available online: https://ec.europa.eu/eurostat/data/database (accessed on 25 January 2019).

Article

Impact of Thermal Radiation and Heat Source/Sink on MHD Time-Dependent Thin-Film Flow of Oldroyed-B, Maxwell, and Jeffry Fluids over a Stretching Surface

Abdul Samad Khan [1,*], Yufeng Nie [1] and Zahir Shah [2]

[1] Department of Applied Mathematics, School of Science, Northwestern Polytechnical University, Dongxiang Road, Chang'an District, Xi'an 710129, China; yfnie@nwpu.edu.cn
[2] Department of Mathematics, Abdul Wali Khan University, Mardan 32300, KP, Pakistan; zahir1987@yahoo.com
* Correspondence: abdulsamadkhan17@mail.nwpu.edu.cn

Received: 1 March 2019; Accepted: 27 March 2019; Published: 2 April 2019

Abstract: In this study paper, we examined the magnetohydrodynamic (MHD) flow of three combined fluids, Maxwell, Jeffry, and Oldroyed- B fluids, with variable heat transmission under the influence of thermal radiation embedded in a permeable medium over a time-dependent stretching sheet. The fluid flow of liquid films was assumed in two dimensions. The fundamental leading equations were changed to a set of differential nonlinear and coupled equations. For this conversion, suitable similarity variables were used. An optimal tactic was used to acquire the solution of the modeled problems. The convergence of the technique has been shown numerically. The obtained analytical and numerical consequences are associated graphically and tabulated. An excellent agreement was obtained between the homotropy analysis method (HAM) and numerical methods. The variation of the skin friction and Nusslet number and their influence on the temperature and concentration profiles were scrutinized. The influence of the thermal radiation, unsteadiness effect, and MHD were the main focus of this study. Furthermore, for conception to be physically demonstrated, the entrenched parameters are discussed graphically in detail along with their effect on liquid film flow.

Keywords: Jeffrey, Maxwell, Oldroyed-B fluids; unsteady stretching surface; magnetic field; homotropy analysis method (HAM)

---

1. Introduction

In the last few years, thin film flow problems have received great attention because of the importance of thin film flows in various technologies. The investigation of thin film flow has obtained importance by its vast application and uses in engineering, technology, and industries. Cable, fiber undercoat, striating of foodstuff, extrusion of metal and polymer, constant forming, fluidization of the device, elastic sheets drawing, chemical treating tools, and exchanges are several uses. In surveillance of these applications, researchers have paid attention to cultivating the examination of liquid film on stretching surface. Emslie et al. [1] investigated thin film flow with applications. During the disk rotating process, they considered the balance between centrifugal and viscous forces. They simplified the Navier–Stokes equations and concluded that the film is uniformly maintained with its continuous thinning property. Higgins [2] considered the influence of the film inertia over a rotating disk. A liquid film over a rotating plate with constant angular velocity was analyzed by Dorfman [3]. The fluid film rotation on an accelerating disk was analyzed by Wang et al. [4]. For the thin and thick film parameter and small accelerating parameters, the asymptotic solutions were obtained. Andersson et al. [5] asymptotically and numerically examined the magnetohydrodynamics (MHD) liquid thin film due on a rotating disk. Over a rotating disk, Dandapat and Singh [6] examined the two layer film flow. The heat transfer flow of thin film flow of nonfluids was deliberated on by Sandeep et al. [7]. Recently, researchers [8–15] examined non-Newtonian nanofluid thin films using different models in different geometries and obtained useful results.

Jeffrey, Maxwell, and Oldroyed-B nanofluids have certain importance in the area of fluid mechanics because of the stress relaxation possessions. Kartini et al. [16] studied the two-dimensional MHD flow and heat transfer of Jeffrey nanofluid over an exponentially stretched plate. Hayat et al., [17] studied the convection of heat transfer in two-dimensional flow of Oldroyed-B nanofluid over a stretching sheet under radiation, and they also found that, with the escalation in radiation parameter, the temperature profile of the nanofluid escalates. Raju et al. [18] debated the presence of a homogenous–heterogeneous reaction in the nonlinear thermal radiation effect on Jeffrey nanofluid flow. Hayat et al. [19] presented a series solution of MHD flow of Maxwell nanofluid over a permeable stretching sheet with suction and injection impacts. Sandeep et al. [20] studied the unsteady mixed convection flow of micropolar fluid over a stretching surface with a non-uniform heat source. Nadeem et al. [21] discussed the heat and mass transfer in peristaltic motion of Jeffrey nanofluid in an annulus. Sheikholeslami [22,23] discussed the hydrothermal behavior of nanofluids flow due to external heated plates. Shah et al. [24–29] investigated MHD nanofluid flow heat transfer, and they mathematically analyzed it with Darcy–Forchheimer phenomena and Cattaneo–Christov flux impacts. Nanofluid flow was studied between parallel plates and stretching sheets. They used analytical and numerical approaches and obtained excellent results. Dawar et al. [30] analyzed the MHD carbon nanotubes (CNTs) Casson nanofluid in rotating channels. Khan et al. [31] studied the 3-D Williamson nanofluid flow over a linear stretching surface.

Lee [32] was the first to investigate the flow with a high Reynolds number (boundary layer concept) over a thin body with variable thickness. In engineering science, the variable thickness extending sheet has many more applications compared to the flat sheet. For instance, for a straightly-extending sheet of an incompressible material, if the extending speed is directly relative to the separation from the slot, the sheet thickness diminishes straightly with the separation. For different materials with various extensibility, the plate/sheet thickness may change as indicated by different profiles. An extending plate/sheet with variable thickness is more realistic. Peristaltic flow of two layers of power law fluids was studied in Usha et al. [33]. Eegunjobi and Makinde [34] mutually derived the effect of buoyancy force and Navier slip in a channel of vertical pores on entropy generation. Numerical analyses of the buoyancy force's influence on the unsteady flow of hydromagnetic by a porous channel with injection/suction were presented by Makinde et al. [35].

Many mathematical problems in the field of engineering and science are complicated, and it is often impossible to solve these types of problems exactly. To find an approximate solution, numerical and analytical techniques are widely implemented in the literature. Liao in 1992 [36,37] investigated the solution of such types of problems by implementing a new proposed technique. This technique was named the homotropy analysis method. He further discussed the convergence of the newly implemented method. A solution is a function of a single variable in the form of a series. Due to fast convergence and strong results, many researchers [38–45] used the homotropy analysis method. Another powerful analytical tool to solve differential equations is the Lie algebra method [46–48].

The goal of the present study was to inspect the MHD flow of three combined nanofluids (Maxwell, Oldroyed-B, and Jeffrey) over a linear stretching surface. The fundamental leading equations were changed to a set of differential nonlinear equations with the support of appropriate correspondence variables. An optimal tactic was used to achieve the solution of the modeled equations, which is nonlinear. The effects of all embedding parameters were studied graphically. Boundary layer methodology was used in the mathematical expansion. The influences of the skin friction, Nusselt number, and Sherwood number on the velocity profile, temperature profile, and concentration profile, respectively, were studied.

2. Problem Formulation

Consider a time-dependent and electric-conducting thin film flow of magnetohydrodynamic Jeffrey, Maxwell, and Oldroyed-B thin film liquids while extending a plate. The flexible sheet starts from an inhibiting slit, which is immovable at the descent of the accommodating system. The Cartesian accommodating system oxyz is adjusted in a well-known manner where ox is equivalent to the plate and oy is smooth to the sheet. The surface of the flow is stretched, applying two equivalent and reversed forces along the x-axis and keeping the origin stationary. The x-axis is taking along the extending surface with stressed velocity $U_w(x,t) = \frac{bx}{1-at}$, in which a, b are constant and the y-axis is perpendicular to it. $T_w(x,t) = T_0 + T_r\left(\frac{bx^2}{2v_f}\right)(1-at)^{-1.5}$ is the wall temperature of the fluid, v_f represents the kinematic viscidness of the fluid, and T_0, T_r are slits. An exterior magnetic ground $B(t) = B_0(1-at)^{-0.5}$ is given normally to the extending sheet (as shown in Figure 1). All the body forces are ignored in the flow field.

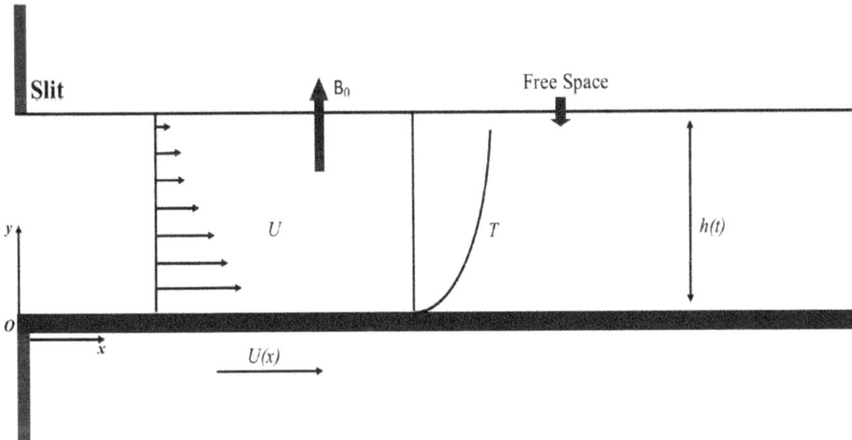

Figure 1. Geometry of the problem.

Considering these hypotheses, the continuity equation, the elementary boundary governing equation, and the heat transfer and concentration equations can be stated as:

$$u_x + u_y = 0 \tag{1}$$

$$\rho_f\left[u_t + uu_x + vu_y + \lambda_1\left(u^2 u_{xx} + v^2 u_{yy} + 2uvu_{xy}\right)\right] = \frac{\mu_f(T)}{1+\lambda_2}\left[u_{yy} + \lambda_3\left(u_{tyy} + uu_{xyy} + vu_{yyy} - u_x u_{yy} + u_y u_{xy}\right)\right] - \sigma_{nf}B^2(t)u, \tag{2}$$

$$T_t + uT_x + vT_y = \frac{1}{\rho c_p}T_{yy} + \frac{Q}{\rho c_p}T_s - T_0 - \frac{1}{\rho c_p}q_y, \tag{3}$$

The applicable boundary conditions are:

$$u = U_w, \ v = 0, \ T = T_w \ \text{at} \ y = 0, \tag{4}$$

$$u_x = 0, T_x = 0, \ \text{at} \ y = h \tag{5}$$

where ϑ_{rd} is:

$$\vartheta_{rd} = -\frac{16\varphi}{3\hat{k}}T_y^4 \tag{6}$$

Applying Taylor series expansion on T^4, we get:

$$T^4 = T_0{}^4 + 4T_0^3(T - T_0)^2 + \dots \tag{7}$$

By ignoring higher-order terms:

$$T^4 = 4TT_0{}^3 - 3T_0{}^4 \tag{8}$$

on which u, v are the velocity constituents alongside the x and y-axes, respectively, σ is the electric conductive parameter, μ, ρ_f, v_f are the dynamic viscosity, density, and kinematic viscosity, respectively and $\gamma_1, \gamma_2, \gamma_3$ represents the relaxation and retardation time ratios. $B(t)$ characterizes the applied magnetics pitch, T_s represents the temperature of the fluid, and T_w characterizes the temperature. Equation (2) manages distinctive fluid models dependent on the below conditions [7]:

I.   The fluid should be Jeffrey fluid when $\lambda_1 = 0$, $\lambda_2 \neq 0$, $\lambda_3 \neq 0$;
II. The fluid should be Maxwell fluid when $\lambda_1 \neq 0$, $\lambda_2 = 0$, $\lambda_3 = 0$;
III. The fluid should be Oldroyed-B fluid when $\lambda_1 \neq 0$, $\lambda_3 \neq 0$ $\lambda_2 = 0$.

Considering the above similarity transformations:

$$\psi = x\sqrt{\frac{vb}{1-at}}f(\eta), u = \psi_y = \frac{bx}{(1-at)}f'(\eta),$$
$$v = -\psi_x = -\sqrt{\frac{vb}{1-at}}f(\eta), \eta = \sqrt{\frac{b}{v(1-at)}}y, y = \sqrt{\frac{v(1-at)}{b}} \tag{9}$$
$$\phi(\eta) = \frac{T-T_0}{T_w-T_0},$$

where prime specifies the derivative with respect to η and ψ indicates the stream function, $h(t)$ specifies the liquid film thickness, and $v = \frac{\mu}{\rho}$ is the kinematics viscosity. The dimensionless film thickness $\beta = \sqrt{\frac{b}{v(1-at)}}h(t)$, which gives $h_t = -\frac{\alpha\beta}{2}\sqrt{\frac{v}{b(1-at)}}$. Substituting Equation (9) into Equations (1)–(6) gives L

$$\frac{1}{(1+\lambda_2)}\left[f''' + \lambda_3\left\{(f'')^2 - ff^{iv} + S\left(2f''' + \frac{\eta}{2}f^{iv}\right) - Mf'\right\}\right] +$$
$$(1 + \varepsilon\kappa)\left[ff'' - (f')^2 - S\left(2f' + \frac{\eta}{2}f''\right) - \lambda_1\left((f)^2 f''' - 2ff'f''\right)\right] = 0 \tag{10}$$

$$(1 + Rd)\theta'' - Pr\left\{\frac{S}{2}(3\theta' + \eta\theta') + (2f' - q)\theta - \theta'f\right\} = 0$$

The boundary constraints are:

$$f(0) = 0, f'(0) = 1, \theta(0) = 1,$$
$$f(\beta) = \frac{S\beta}{2}, f''(\beta) = 0, \theta'(\beta) = 0$$

Skin friction is defined as $C_f = \dfrac{\left(\widetilde{S}_{xy}\right)_{y=0}}{\rho \tilde{u}_w^2}$,

$$C_f = \frac{1+\lambda_1}{1+\lambda_2}f''(0)$$

The Nusselt number is:

$$u = \frac{Q_w}{\hat{k}(T - T_h)}, \tag{11}$$

$$u = \frac{1}{2}(1 - \alpha t)^{-\frac{1}{2}} - \theta'(0)Re^{-\frac{1}{2}}f_x,$$

After generalization, the physical constraints are as follows: $\lambda_1, \lambda_2, \lambda_3$ are the parameters of relaxation and retardation ratios, $S = \frac{a}{b}$ is the measure of unsteadiness of the nondimensional, $M = \frac{\sigma_f B_0^2}{b \rho_f}$ denotes the magnetic field parameter, and $Pr = \frac{\rho v c_p}{k}$ denotes the Prandtl number.

3. Solution by HAM

Equations (8)–(10), considering the boundary condition in Equation (11), are solved with HAM. The solutions encircled the auxiliary parameter h, which normalizes and switches to a conjunction of the solutions.

The following is the initial guess:

$$f_0(\eta) = \frac{(2-S)\eta^3 + (3S-6)\beta\eta^2 + 4\beta^2\eta}{4\beta^2}, \ \theta(\eta) = 1 \tag{12}$$

L_f, L_θ reperesent linar operators:

$$L_f(f) = f^{iv}, \ L_\theta(\theta) = \theta' \tag{13}$$

where:

$$L_f\left(\frac{e_1}{6}\eta^3 + \frac{e_2}{2}\eta^2 + e_3\eta + e_4\right) = 0, \ L_\theta(e_5 + e_6\eta) = 0 \tag{14}$$

Non-linear operators N_f, N_θ are given as:

$$N_f\left[f(\eta; r)\right] = \frac{1}{(1+\lambda_2)}\left[\begin{array}{c} \frac{\partial^3 f(\eta;r)}{\partial\eta^3} + \lambda_3\left(\frac{\partial^2 f(\eta;r)}{\partial\eta^2}\right)^2 - f(\eta;r)\frac{\partial^4 f(\eta;r)}{\partial\eta^4} + \\ S\left(2\frac{\partial^3 f(\eta;r)}{\partial\eta^3} + \frac{\eta}{2}\frac{\partial^4 f(\eta;r)}{\partial\eta^4}\right) - M\frac{\partial f(\eta;r)}{\partial\eta} \end{array}\right] +$$
$$(1+\varepsilon\kappa)\left[\begin{array}{c} f(\eta;r)\frac{\partial^2 f(\eta;r)}{\partial\eta^2} - \left(\frac{\partial f(\eta;r)}{\partial\eta}\right)^2 - S\left(2\frac{\partial f(\eta;r)}{\partial\eta} + \frac{\eta}{2}\frac{\partial^2 f(\eta;r)}{\partial\eta^2}\right) - \\ \lambda_1\left((f(\eta;r))^2\frac{\partial^3 f(\eta;r)}{\partial\eta^3} - 2f(\eta;r)\frac{\partial f(\eta;r)}{\partial\eta}\frac{\partial^2 f(\eta;r)}{\partial\eta^2}\right) \end{array}\right] \tag{15}$$

$$\begin{aligned} N_\theta[&f(\eta;r), \theta(\eta;r)] \\ &= (1+Rd)\frac{\partial^2\theta(\eta;r)}{\partial\eta^2} \\ &\quad -Pr\left\{\begin{array}{c} \frac{S}{2}\left(3\theta(\eta;r) + \eta\frac{\partial\theta(\eta;r)}{\partial\eta}\right) + \\ \left(2\frac{\partial f(\eta;r)}{\partial\eta} - q\right)\theta(\eta;r) - f(\eta;r)\frac{\partial\theta(\eta;r)}{\partial\eta} \end{array}\right\} \end{aligned} \tag{16}$$

The 0th-order problem is defined as:

$$L_f[f(\eta;r) - f_0(\eta)](1-r) - r_f N_f[f(\eta;r)] \tag{17}$$

$$L_\theta[\theta(\eta;r) - \theta_0(\eta)](1-r) = r_\theta N_\theta[f(\eta;r), \theta(\eta;r)] \tag{18}$$

The correspondent boundary constraints are:

$$\begin{aligned} &f(\eta;r)\big|_{\eta=0} = 0, \ \frac{\partial f(\eta;r)}{\partial\zeta}\Big|_{\eta=0} = 1, \ \frac{\partial^2 f(\zeta;r)}{\partial\eta^2}\Big|_{\eta=\beta} = 0, \ f(\eta;r)\big|_{\eta=\beta} = 0 \\ &\theta(\eta;r)\big|_{\eta=0} = 1, \ \frac{\partial\theta(\eta;r)}{\partial\eta}\Big|_{\eta=\beta} = 0, \end{aligned} \tag{19}$$

In the case of $r = 0$, $r = 1$:

$$f(\eta;1) = f(\eta), \ \theta(\eta;1) = \theta(\eta) \tag{20}$$

After applying Taylor's series:

$$f(\eta; r) = f_0(\eta) + \sum_{n=1}^{\infty} f_n(\eta) r^n,$$
$$\theta(\eta; r) = \theta_0(\eta) + \sum_{n=1}^{\infty} \theta_n(\eta) r^n, \tag{21}$$

where:

$$f_n(\eta) = \frac{1}{n!} \left. \frac{\partial f(\eta; r)}{\partial \eta} \right|_{r=0}, \quad \theta_n(\eta) = \frac{1}{n!} \left. \frac{\partial \theta(\eta; r)}{\partial \eta} \right|_{r=0} \tag{22}$$

The subordinate limitations $\hbar_f$, $\hbar_\theta$ are selected such that the series in Equation (23) converges at $r = 1$. Switching $r = 1$ in Equation (23), we obtain:

$$f(\eta) = f_0(\eta) + \sum_{n=1}^{\infty} f_n(\eta),$$
$$\theta(\eta) = \theta_0(\eta) + \sum_{n=1}^{\infty} \theta_n(\eta), \tag{23}$$

The nth-order problem:

$$L_f \left[f_n(\eta) - \chi_n f_{n-1}(\eta) \right] = \hbar_f \, R_n^f(\eta),$$
$$L_\theta \left[\theta_n(\eta) - \chi_n \theta_{n-1}(\eta) \right] = \hbar_\theta \, R_n^\theta(\eta), \tag{24}$$

The consistent boundary conditions are:

$$f_n(0) = f_n'(0) = \theta_n(0) = 0$$
$$f_n(\beta) = f_n''(\beta) = \theta_n'(\beta) = 0 \tag{25}$$

and:

$$\chi_n = \begin{cases} 0, & \text{if } r \leq 1 \\ 1, & \text{if } r > 1 \end{cases} \tag{26}$$

HAM Solution Convergence

In this section, the convergence of the problem of the modeled and solved equations is discussed. Figure 2 presents the *h*-curves of the combined temperature and velocity functions. A valid region of convergent is observed in Figure 2. In Table 1, numerical convergence is presented. A fast convergence of HAM is shown in Table 1.

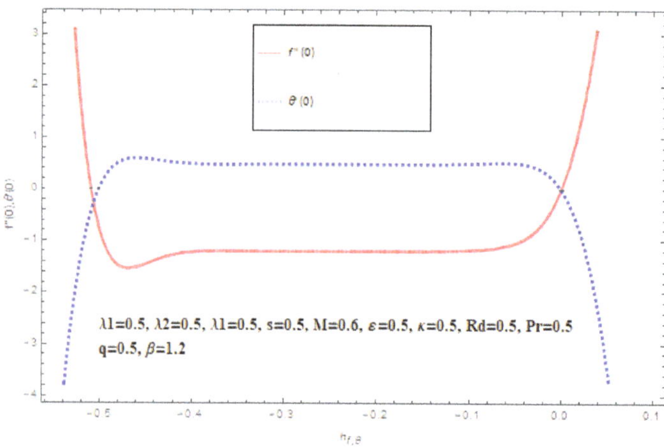

Figure 2. The combined h-curve graph of the velocity and temperature profiles.

Table 1. Fast convergence of HAM (homotropy analysis method).

Solution Approximations	$f''(0)$	$\theta'(0)$
1	0.448400	0.122519
4	0.858574	−0.282846
7	0.912994	−0.356251
10	0.920043	−0.369590
13	0.920946	−0.372026
16	0.921062	−0.372472
19	0.921077	−0.372554
22	0.921080	−0.372569
25	0.921080	−0.372572
28	0.921080	−0.372572

4. Results and Discussion

This work investigated thin particle flow (Maxwell, Jeffry, and Oldroyed-B) fluids considering the influence of MHD and radiation with a time-dependent porous extending surface. This subsection presents the physical impact of dissimilar implanting parameters over velocity distributions $f(\eta)$ and temperature distribution $\Theta(\eta)$ as exemplified in Figures 2–12.

Figure 3 shows the effect of κ on velocity field $f(\eta)$. There is a direct relation between the cohesive and adhesive forces and viscosity. The overall fluid motion is reduced due to the rising in κ, which generates strong cohesive and adhesive forces that cause resistance.

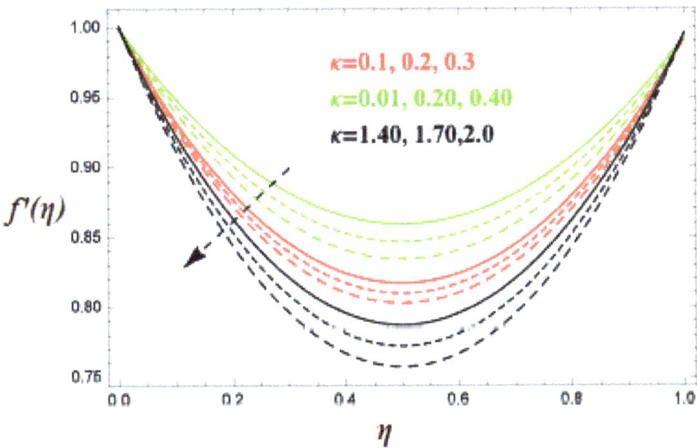

Figure 3. Variations in velocity field $f(\eta)$ for different values of κ.

Figure 4 shows the effect of β during the motion of the flow. Increasing β decreases the flow velocity, as it is a reducing function of the velocity and thickness of the liquid film. Substantially, the viscid forces rising with higher numbers of β cause the fluid motion. Thus, fluid velocity drops.

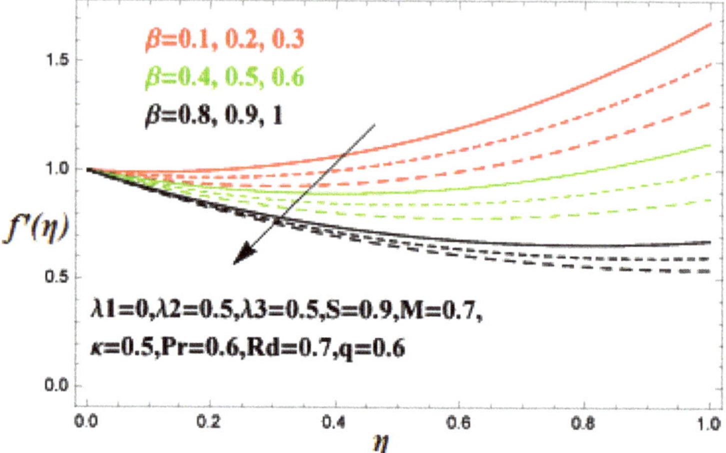

Figure 4. Fluctuations in velocity field $f(\eta)$ for various values of β.

Figure 5 shows the impact of thin film thickness on the temperature profile. It is observed that an increase in β decreases temperature profile $\theta(\eta)$.

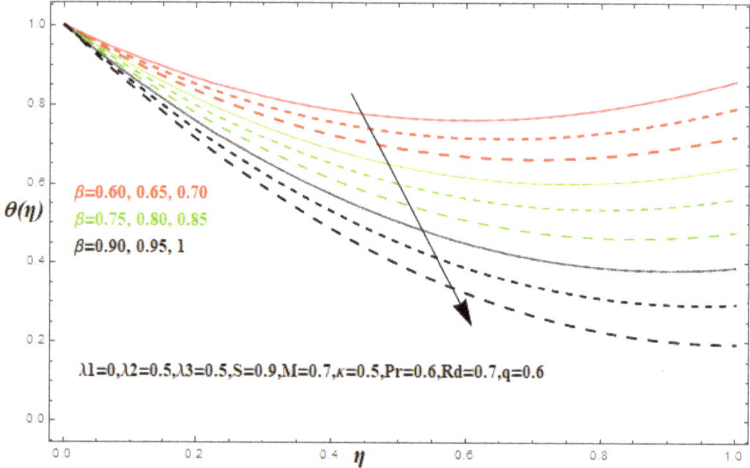

Figure 5. Variations in the temperature gradient $\theta(\eta)$ for the various measures of β.

Figure 6 describes the features of M on (η). The velocity profile reduces due to the rise in M. The object of this phenomenon is Lorentz force. This force is responsible for the reduction in the motion of fluid.

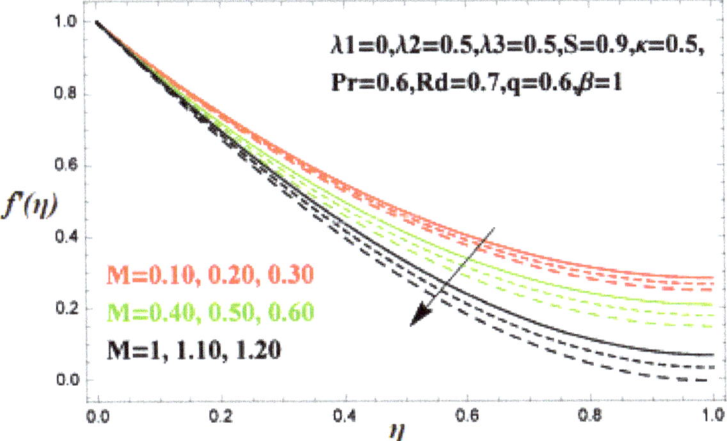

Figure 6. Dissimilarities in $f(\eta)$ for different numbers of M.

Figure 7 shows that with different values of M, the concentration field increases because an increase in oblique magnetic pitch on the fluid augments the Lorentz force. Lorentz force is the combination of electric and magnetic force on a point charge due to electromagnetic fields. This force is lower at the center of the channel and at the boundaries, but up near the walls. Velocity increases with the unsteady constraint S.

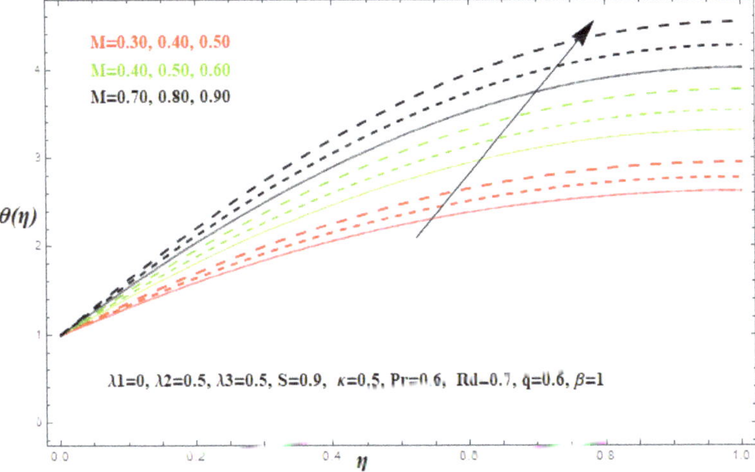

Figure 7. Changes in temperature gradient $\theta(\eta)$ for diverse measures of M.

Figure 8 demonstrates the performance of the unsteadiness constraint S on $f(\eta)$. It is concluded that $f(\eta)$ is directly proportional to S. The fluid motion increases due to an increase in S. Thus, the solution is dependent on S, S.

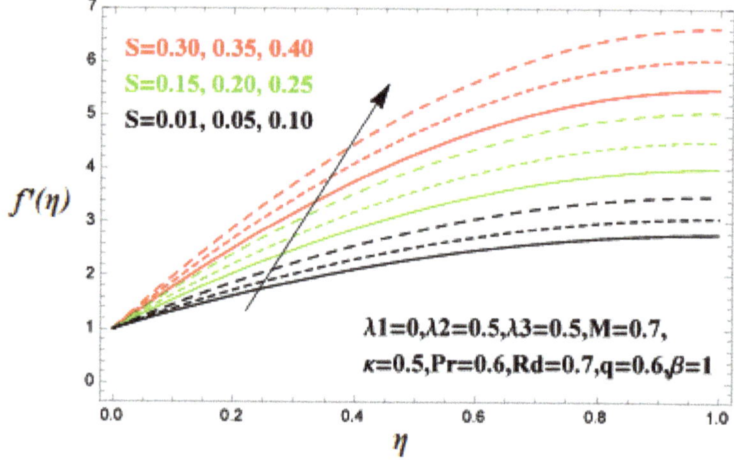

Figure 8. Fluctuations in $f(\eta)$ for several numbers of S.

Figure 9 shows the effect of S on the heat profile $\theta(\eta)$. It was concluded that $\theta(\eta)$ is directly proportional to S. An increase in the unsteadiness parameter S increases the temperature which, in turn, augments the kinetic energy of the fluid flow; thus, the thin film moment increases.

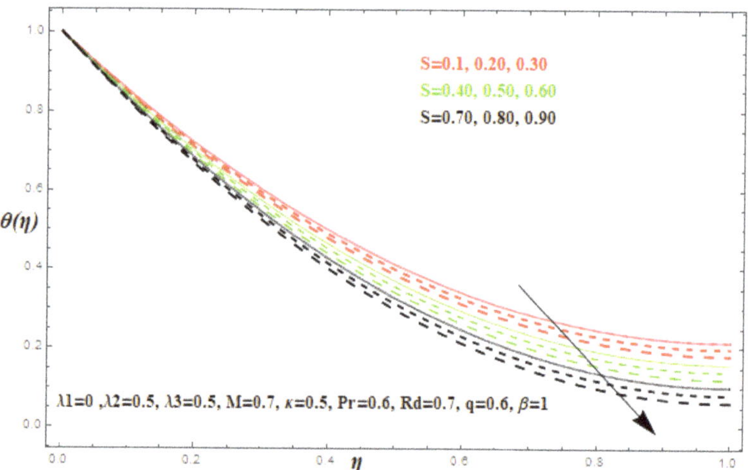

Figure 9. Changes in temperature gradient $\theta(\eta)$ for diverse measures of S.

Figure 10 shows the effect of q. Increasing q increases the temperature field. Actually, raising q increases the kinetic energy of thin film, which results in increasing the internal heat.

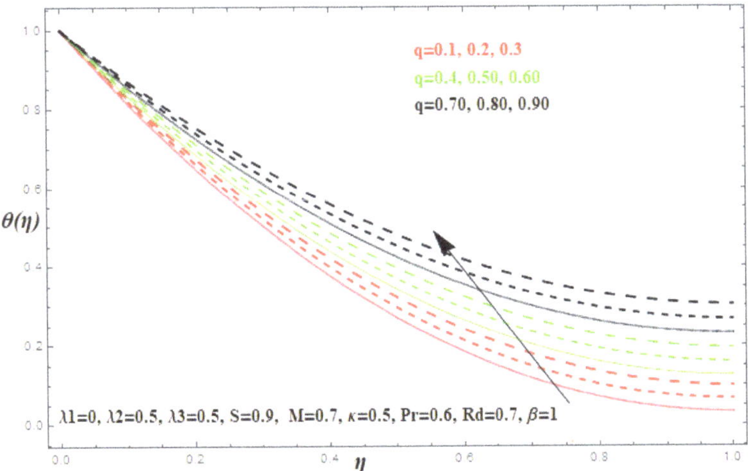

Figure 10. Changes in temperature gradient $\theta(\eta)$ for diverse measures of q.

The influence of radioactivity parameter Rd on $\theta(\eta)$ is shown in Figure 11. Radioactivity is an important part on the inclusive surface heat transmission when the coefficient of convection heat transmission is small. When thermal radiation is increased, the thermal radiation augments the temperature in the boundary layer area in the fluid layer. This increase leads to a drop in the rate of cooling for liquid film flow.

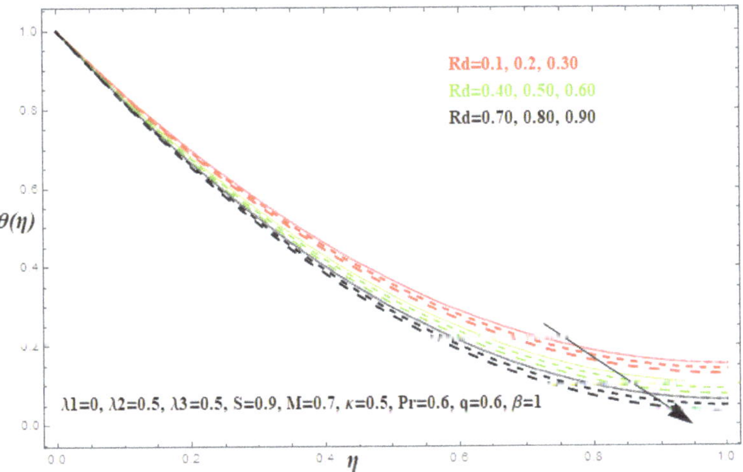

Figure 11. Changes in temperature gradient $\theta(\eta)$ for diverse measures of Rd.

The influence of Pr on the temperature distributions $\theta(\eta)$ is shown in Figure 12. The temperature distribution varies inversely with Pr. Clearly, the temperature distribution decreases for large values of Pr and increases for small values of Pr. Physically, the fluids having a small Pr have larger thermal diffusivity and vice versa. Thus, large Pr causes the thermal boundary layer to decrease.

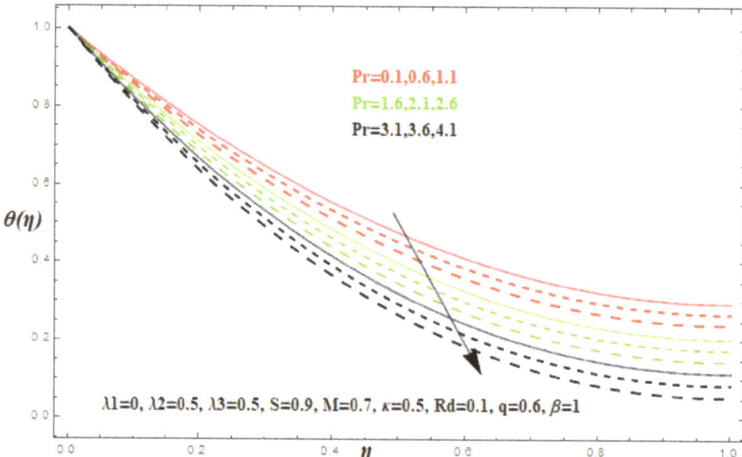

Figure 12. Changes in temperature gradient $\theta(\eta)$ for diverse measures of Pr.

Figures 13 and 14 illustrate the comparison of HAM and numerical solution for velocities and temperature functions. An excellent agreement is found here.

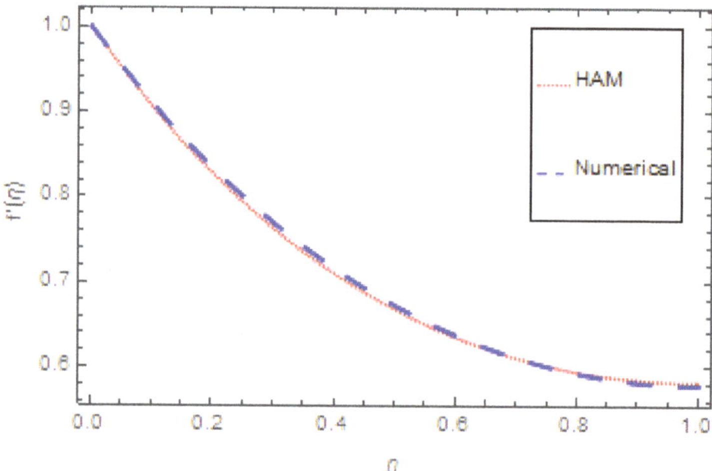

Figure 13. Comparison graph of HAM and numerical solution for $f(\eta)$.

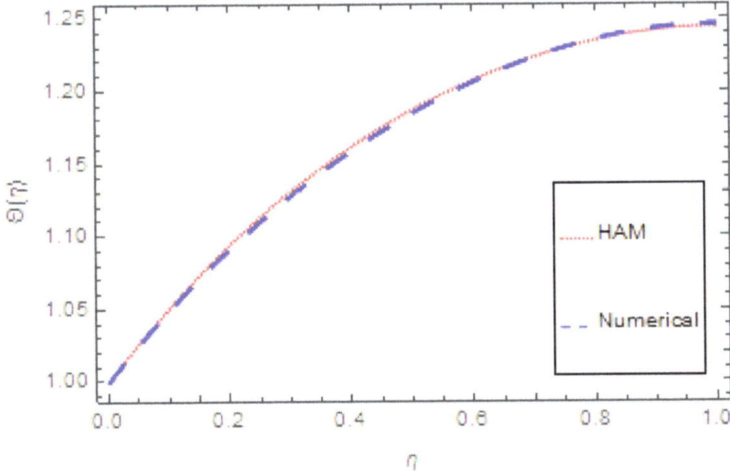

Figure 14. Comparison graph of HAM and numerical solution for $\theta(\eta)$.

Tables 2 and 3 present a comparison of HAM and numerical solution for velocities and temperature functions. We have observed an excellent covenant between the analytical and numerical approach.

The numerical values of surface temperature $\theta(\beta)$ for different values of M, Rd and k are given in Table 2. Increasing values of M, Rd, and k increase surface temperature $\theta(\beta)$, whereas the opposite effect was found for Pr, that is, large values of Pr reduce surface temperature $\theta(\beta)$.

Table 2. Comparison of HAM and numerical solution for $f(\eta)$.

η	HAM Solution $f(\eta)$	Numerical Solution $f(\eta)$	Absolute Error
0.0	1.00000	1.00000	0.00000
0.1	0.907327	0.911301	3.9×10^{-3}
0.2	0.829165	0.835332	6.2×10^{-3}
0.3	0.764141	0.770975	6.8×10^{-3}
0.4	0.710984	0.717248	6.3×10^{-3}
0.5	0.668524	0.673301	4.8×10^{-3}
0.6	0.635691	0.638412	2.7×10^{-3}
0.7	0.611519	0.611979	4.5×10^{-4}
0.8	0.595141	0.593511	1.8×10^{-3}
0.9	0.585791	0.582627	7.7×10^{-3}
1.0	0.582805	0.579045	3.8×10^{-3}

Table 3. Comparison of HAM and numerical solution for $\theta(\eta)$.

η	HAM Solution $\theta(\eta)$	Numerical Solution $\theta(\eta)$	Absolute Error
0.0	1.0000	1.00000	0.00000
0.1	1.05078	1.04882	1.9×10^{-3}
0.2	1.09412	1.09106	3.1×10^{-3}
0.3	1.13081	1.12742	3.3×10^{-3}
0.4	1.16156	1.15845	3.1×10^{-3}
0.5	1.18696	1.18458	2.3×10^{-3}
0.6	1.20745	1.20608	1.3×10^{-3}
0.7	1.22335	1.22308	2.7×10^{-4}
0.8	1.23481	1.23555	7.3×10^{-3}
0.9	1.24184	1.24331	1.4×10^{-3}
1.0	1.24425	1.2460	1.7×10^{-3}

The numerical value of the gradient of wall temperature $\theta'(0)$ for different values of embedded parameters Rd, β, Pr, S is shown in Tables 4 and 5. Larger values of thermal radiation Rd, β, and Pr decrease the wall temperature, whereas S increases the wall temperature gradient $\theta'(0)$.

Table 4. Numerical values for the skin friction coefficient for several physical parameters when: $\lambda_1 = 0.1, \lambda_2 = 1.0, \lambda_3 = 0.1, S = 0.5, M = 0.6, \kappa = 0.7, Pr = 1.0, \beta = 1$.

S	κ	M	β	$\frac{1+\lambda_1}{1+\lambda_2} f''(0)$ Jeffrey Fluid	$\frac{1+\lambda_1}{1+\lambda_2} f''(0)$ Maxwell Fluid	$\frac{1+\lambda_1}{1+\lambda_2} f''(0)$ Oldroyed-B Fluid
0.5	0.5	0.6	1.0	-0.705657	-1.40417	-1.29373
0.6				-0.690279	-1.34383	-1.25575
0.7				-0.676699	-2.19921	-1.23210
0.5	0.5			-0.705657	-1.40417	-1.29373
	0.6			-0.719191	-1.43646	-1.32213
	0.7			-0.732514	-1.46831	-1.35012
	0.5	0.6		-0.705657	-1.40417	-1.29373
		0.7		-0.725666	-1.44510	-1.33013
		0.8		-0.745259	-1.48521	-1.36548
		0.6	1.0	-0.705657	-1.40417	-1.29373
			1.1	-0.725279	-1.469866	-1.33341
			1.2	-0.740967	-1.54394	-0.268147
			1.0	-0.705657	-1.40417	-1.29373

Table 5. Numerical values of the local Nusselt number for several physical parameters, when: $\lambda_1 = 0.1, \lambda_2 = 1.0, \lambda_3 = 0.1, S = 0.5, M = 0.6, \kappa = 0.7, Pr = 1.0, \beta = 1, Rd = 0.5, q = 0.5$.

S	κ	M	β	Rd	q	$\theta'(0)$ Jeffrey Fluid	$\theta'(0)$ Maxwell Fluid	$\theta'(0)$ Oldroyed-B Fluid
0.5	0.5	0.6	1.0	0.5	0.5	−0.550388	−0.590996	−0.616292
0.6						−0.545322	−0.590885	−0.611008
0.7						−0.539440	−0.683523	−0.605811
0.5	0.5					−0.550388	−0.590996	−0.616292
	0.6					−0.543228	−0.582409	−0.608402
	0.7					−0.536265	−0.574031	−0.600680
	0.5	0.6				−0.550388	−0.590996	−0.616292
		0.7				−0.538941	−0.580121	−0.605984
		0.8				−0.527911	−0.569593	−0.595950
		0.6	1.0			−0.550388	−0.590996	−0.616292
			1.1			−0.551698	−0.589585	−0.619001
			1.2			−0.547039	−0.580929	−0.376664
			1.0	0.5		−0.550388	−0.590996	−0.616292
				0.6		−0.517715	−0.555700	−0.579345
				0.7		−0.488677	−0.524350	−0.546542
				0.5	0.5	−0.550388	−0.590996	−0.616292

5. Conclusions

The MHD flow of three combined fluids, Maxwell, Jeffry, and Oldroyed-B fluid, with variable heat transmission under the influence of thermal radiation was investigated. The fundamental leading equations were changed to a set of differential nonlinear and coupled equations. An optimal and numerical tactic was used to acquire the solution of the modeled problems.

The key conclusions of this work are as follows:

- The velocity profile shows an increase with increasing values of the unsteadiness parameter St, while increasing values of the magnetic parameter cause the decline of the velocity profile of the nanofluid film.
- Rd augmented heat flux
- Higher values of Pr increase the surface temperature.
- Fast convergence is shown numerically and graphically.
- The liquid film flow effects on the flow of three fluids (Jeffrey, Maxwell, and Oldroyd-B fluids) are presented using graphs and tables.
- It was found that the coefficient of skin friction rises with the larger rates of the magnetic parameter M and the unsteadiness parameter St, whereas the coefficient of skin friction decreases with the higher values of the stretching and thickness parameters.
- It was found that liquid film flow is affected by the Lorentz's force.
- The obtained analytical and numerical consequences are presented graphically and tabulated. An excellent agreement was obtained between HAM and the numerical method.

Author Contributions: A.S.K. and Z.S. modeled the problem and wrote the manuscript. Y.N. thoroughly checked the mathematical modeling and English corrections. A.S.K. and Z.S. solved the problem using Mathematica software. Y.N. contributed to the results and discussions. All authors finalized the manuscript after its internal evaluation.

Funding: This research was funded by National Natural Science Foundation of China No. [11471262].

Conflicts of Interest: The authors declare no conflict of interest.

Nomenclature

J Current density	α_1, α_2 Material constants
μ Coefficient of viscosity	μ_0 Magnetic permeability
T Cauchy stress tensor	M Magnetic parameter
σ Electric conductivity	α non-Newtonian parameter
E Electric field	Ω Pressure gradient parameter
ω Frequency parameter	A_1 and A_2 Rivlin–Ericksen tensor
p pressure	t Time variable
I Identity tensor chord	Rd Radiation paramete
St Unsteadiness parameter	
Pr Prandtl number	Nu Nusselt number
C_f Skin friction	

References

1. Emslie, A.G.; Bonner, F.T.; Peck, L.G. Flow of a viscous liquid on a rotating disk. *J. Appl. Phys.* **1958**, *5*, 858–862. [CrossRef]
2. Higgins, B.G. Film flow on a rotating disk. *Phy. Fluids.* **1986**, *29*. [CrossRef]
3. Dorfman, I.A. Flow and heat transfer in a film of viscous liquid on a rotating disk. *J. Eng. Phys.* **1967**, *12*, 162–166. [CrossRef]
4. Wang, C.Y.; Watson, L.T.; Alexander, K.A. Spinning of a liquid.lm from an accelerating disc. *IMA J. Appl. Math.* **1991**, *46*, 201–210. [CrossRef]
5. Andersson, H.I.; Holmedal, B.; Dandapat, B.S.; Gupta, A.S. Magnetohydrodynamic melting flow from a horizontal rotating disk. *Math. Models Methods Appl. Sci.* **1993**, *3*. [CrossRef]
6. Dandapat, B.S.; Singh, S.K. Unsteady two-layer film flow on a non-uniform rotating disk in presence of uniform transverse magnetic field. *Appl. Math. Comput.* **2015**, *258*, 545–555. [CrossRef]
7. Sandeep, N.; Malvandi, A. Enhanced heat transfer in liquid thin film flow of non-Newtonian nanofluids embedded with graphene nanoparticles. *Adv. Powder Technolog.* **2016**, *27*, 2448–2456. [CrossRef]
8. Khan, N.; Zuhra, S.; Shah, Z.; Bonyah, E.; Khan, W.; Islam, S. Slip flow of Eyring-Powell nanoliquid film containing graphene nanoparticles. *AIP Adv.* **2018**, *8*. [CrossRef]
9. Ullah, A.; Alzahrani, E.O.; Shah, Z.; Ayaz, M.; Islam, S. Nanofluids Thin Film Flow of Reiner-Philippoff Fluid over an Unstable Stretching Surface with Brownian Motion and Thermophoresis Effects. *Coatings* **2019**, *9*, 21. [CrossRef]
10. Shah, Z.; Bonyah, E.; Islam, S.; Khan, W.; Ishaq, M. Radiative MHD thin film flow of Williamson fluid over an unsteady permeable stretching. *Heliyon* **2018**, *4*. [CrossRef]
11. Ishaq, M.; Ali, G.; Shah, S.I.A.; Shah, Z.; Muhammad, S.; Hussain, S.A. Nanofluid Film Flow of Eyring Powell Fluid with Magneto Hydrodynamic Effect on Unsteady Porous Stretching Sheet. *J. Math.* **2019**, *51*, 131–153.
12. Jawad, M.; Shah, Z.; Islam, S.; Islam, S.; Bonyah, E.; Khan, Z.A. Darcy-Forchheimer flow of MHD nanofluid thin film flow with Joule dissipation and Navier's partial slip. *J. Phys. Commun.* **2018**, *2*. [CrossRef]
13. Jawad, M.; Shah, Z.; Islam, S.; Majdoubi, J.; Tlili, I.; Khan, W.; Khan, I. Impact of Nonlinear Thermal Radiation and the Viscous Dissipation Effect on the Unsteady Three-Dimensional Rotating Flow of Single-Wall Carbon Nanotubes with Aqueous Suspensions. *Symmetry* **2019**, *11*, 207. [CrossRef]
14. Khan, N.S.; Gul, T.; Islam, S.; Khan, A.; Shah, Z. Brownian Motion and Thermophoresis Effects on MHD Mixed Convective Thin Film Second-Grade Nanofluid Flow with Hall Effect and Heat Transfer Past a Stretching Sheet. *J. Nanofluids* **2017**, *6*, 1–18. [CrossRef]
15. Tahir, F.; Gul, T.; Islam, S.; Shah, Z.; Khan, A.; Khan, W.; Ali, L.; Muradullah. Flow of a Nano-Liquid Film of Maxwell Fluid with Thermal Radiation and Magneto Hydrodynamic Properties on an Unstable Stretching Sheet. *J. Nanofluids* **2017**, *6*, 1–10. [CrossRef]
16. Kartini, A.; Zahir, H.; Anuar, I. Mixed convection Jeffrey fluid flow over an exponentially stretching sheet with magnetohydrodynamic effect. *AIP Adv.* **2016**, *6*. [CrossRef]

17. Hayat, T.; Hussain, T.; Shehzad, S.A.; Alsaedi, A. Flow of Oldroyd-B fluid with nano particles and thermal radiation. *Appl. Math. Mech. Engl. Ed.* **2015**, *36*, 69–80. [CrossRef]

18. Raju, C.S.K.; Sandeep, N.; Gnaneswar, R.M. Effect of nonlinear thermal radiation on 3D Jeffrey fluid flow in the presence of homogeneous–heterogeneous reactions. *Int. J. Eng. Res. Afr.* **2016**, *21*, 52–68. [CrossRef]

19. Hayat, T.; Abbas, Z.; Sajid, M. Series solution for the upper convected Maxwell fluid over a porous stretching plate. *Phys. Lett. A* **2006**, *358*, 396–403. [CrossRef]

20. Sandeep, N.; Sulochana, C. Dual solutions for unsteady mixed convection flow of MHD micropolar fluid over a stretching/ shrinking sheet with non-uniform heat source/sink. *Eng. Sci. Technol. Int. J.* **2015**, *18*. [CrossRef]

21. Nadeem, S.; Akbar, N.S. Influence of haet and mass transfer on a peristaltic motion of a Jeffrey-six constant fluid in an annulus. *Heat Mass. Transfer.* **2010**, *46*, 485–493. [CrossRef]

22. Sheikholeslami, M.; Shah, Z.; Shafi, A.; Khan, I.; Itili, I. Uniform magnetic force impact on water based nanofluid thermal behavior in a porous enclosure with ellipse shaped obstacle. *Sci. Rep.* **2019**, *9*, 1196. [CrossRef] [PubMed]

23. Sheikholeslami, M.; Shah, Z.; Tassaddiq, A.; Shafee, A.; Khan, I. Application of Electric Field for Augmentation of Ferrofluid Heat Transfer in an Enclosure Including Double Moving Walls. *IEEE Access* **2019**, *7*, 21048–21056. [CrossRef]

24. Shah, Z.; Dawar, A.; Islam, S.; Khan, I.; Ching, D.L.C. Darcy-Forchheimer Flow of Radiative Carbon Nanotubes with Microstructure and Inertial Characteristics in the Rotating Frame. *Case Stud. Thermal Eng.* **2018**, *12*, 823–832. [CrossRef]

25. Shah, Z.; Islam, S.; Ayaz, H.; Khan, S. Radiative Heat and Mass Transfer Analysis of Micropolar Nanofluid Flow of Casson Fluid between Two Rotating Parallel Plates with Effects of Hall Current. *ASME J. Heat Transf.* **2019**, *141*. [CrossRef]

26. Shah, Z.; Islam, S.; Gul, T.; Bonyah, E.; Altaf Khan, M. The Elcerical MHD And Hall Current Impact On Micropolar Nanofluid Flow Between Rotating Parallel Plates. *Res. Phys.* **2018**, *9*. [CrossRef]

27. Shah, Z.; Bonyah, E.; Islam, S.; Gul, T. Impact of thermal radiation on electrical mhd rotating flow of carbon nanotubes over a stretching sheet. *AIP Adv.* **2019**, *9*. [CrossRef]

28. Shah, Z.; Dawar, A.; Islam, S.; Khan, I.; Ching, D.L.C.; Khan, A.Z. Cattaneo-Christov model for Electrical MagnetiteMicropoler Casson Ferrofluid over a stretching/shrinking sheet using effective thermal conductivity model. *Case Stud. Therm. Eng.* **2018**, *13*, 100352. [CrossRef]

29. Shah, Z.; Tassaddiq, A.; Islam, S.; Alklaibi, A.; Khan, I. Cattaneo–Christov Heat Flux Model for Three-Dimensional Rotating Flow of SWCNT and MWCNT Nanofluid with Darcy–Forchheimer Porous Medium Induced by a Linearly Stretchable Surface. *Symmetry* **2019**, *11*, 331. [CrossRef]

30. Dawar, A.; Shah, Z.; Islam, S.; Idress, M.; Khan, W. Magnetohydrodynamic CNTs Casson Nanofl uid and Radiative heat transfer in a Rotating Channels. *J. Phys. Res. Appl.* **2018**, *1*, 17–32.

31. Khan, A.S.; Nie, Y.; Shah, Z.; Dawar, A.; Khan, W.; Islam, S. Three-Dimensional Nanofluid Flow with Heat and Mass Transfer Analysis over a Linear Stretching Surface with Convective Boundary Conditions. *Appl. Sci.* **2018**, *8*, 2244. [CrossRef]

32. Lee, L.L. Boundary layer over a thin needle. *Phys. Fluids.* **1967**, *10*, 822–828. [CrossRef]

33. Usha, S., Ramachandra, R.R. Peristaltic transport of two-layered power-law fluids. *J. Biomech. Eng.* **1997**, *119*, 483–488. [CrossRef]

34. Eegunjobi, A.S.; Makinde, O.D. Combined effect of buoyancy force and Navier slip on entropy generation in a vertical porous channel. *Entropy* **2012**, *14*, 1028–1044. [CrossRef]

35. Makinde, O.D.; Chinyoka, T. Numerical investigation of buoyancy effects on hydromagnetic unsteady flow through a porous channel with suction/injection. *Mech. Sci. Technol.* **2013**, *27*, 1557–1568. [CrossRef]

36. Liao, S.J. An explicit totally analytic approximate solution for Blasius viscous flow problems. *Int. J. Non-Linear Mech.* **1999**, *34*, 759–778. [CrossRef]

37. Liao, S.J. On the analytic solution of Magneto hydrodynamic flows of non-Newtonian fluids over a stretching sheet. *J. Fluid Mech.* **2003**, *488*, 189–212. [CrossRef]

38. Nasir, S.; Shah, Z.; Islam, S.; Khan, W.; Khan, S.N. Radiative flow of magneto hydrodynamics single-walled carbon nanotube over a convectively heated stretchable rotating disk with velocity slip effect. *Adv. Mech. Eng.* **2019**, *11*, 1–11. [CrossRef]

39. Nasir, S.S.; Shah, Z.; Islam, S.; Khan, W.; Bonyah, E.; Ayaz, M.; Khan, A. Three dimensional Darcy-Forchheimer radiated flow of single and multiwall carbon nanotubes over a rotating stretchable disk with convective heat generation and absorption. *AIP Adv.* **2019**, *9*. [CrossRef]

40. Khan, A.; Shah, Z.; Islam, S.; Dawar, A.; Bonyah, E.; Ullah, H.; Khan, Z.A. Darcy-Forchheimer flow of MHD CNTs nanofluid radiative thermal behaviour andconvective non uniform heat source/sink in the rotating frame with microstructureand inertial characteristics. *AIP Adv.* **2018**, *8*. [CrossRef]

41. Khan, A.; Shah, Z.; Islam, S.; Khan, S.; Khan, W.; Khan, Z.A. Darcy–Forchheimer flow of micropolar nanofluid between two plates in the rotating frame with non-uniform heat generation/absorption. *Adv. Mech. Eng.* **2018**, *10*, 1–16. [CrossRef]

42. Feroz, N.; Shah, Z.; Islam, S.; Alzahrani, E.O.; Khan, W. Entropy Generation of Carbon Nanotubes Flow in a Rotating Channel with Hall and Ion-Slip Effect Using Effective Thermal Conductivity Model. *Entropy* **2019**, *21*, 52. [CrossRef]

43. Alharbi, S.O.; Dawar, A.; Shah, Z.; Khan, W.; Idrees, M.; Islam, S.; Khan, I. Entropy Generation in MHD Eyring–Powell Fluid Flow over an Unsteady Oscillatory Porous Stretching Surface under the Impact of Thermal Radiation and Heat Source/Sink. *Appl. Sci.* **2018**, *8*, 2588. [CrossRef]

44. Muhammad, S.; Ali, G.; Shah, Z.; Islam, S.; Hussain, S.A. The Rotating Flow of Magneto Hydrodynamic Carbon Nanotubes over a Stretching Sheet with the Impact of Non-Linear Thermal Radiation and Heat Generation/Absorption. *Appl. Sci.* **2018**, *8*, 482. [CrossRef]

45. Saeed, A.; Islam, S.; Dawar, A.; Shah, Z.; Kumam, P.; Khan, W. Influence of Cattaneo–Christov Heat Flux on MHD Jeffrey, Maxwell, and Oldroyd-B Nanofluids with Homogeneous-Heterogeneous Reaction. *Symmetry* **2019**, *11*, 439. [CrossRef]

46. Shang, Y. Analytical Solution For An In-Host Viral Infection Model With Time-Inhomogeneous Rates. *Acta Phys. Polon. B* **2015**, *46*, 1567–1577. [CrossRef]

47. Shang, Y. A lie algebra approach to susceptible-infected-susceptible epidemics. *Electr. J. Different. Equat.* **2012**, *233*, 1–7.

48. Shang, Y. Lie algebraic discussion for aflnity based information diffusion in social networks. *Open Phys.* **2017**, *15*, 705–711. [CrossRef]

Article

Impact of Thermal Radiation on Magnetohydrodynamic Unsteady Thin Film Flow of Sisko Fluid over a Stretching Surface

Abdul Samad Khan [1,*], Yufeng Nie [1] and Zahir Shah [2]

[1] Department of Applied Mathematics, School of Science, Northwestern Polytechnical University, Dongxiang Road, Chang'an District, Xi'an 710129, China; yfnie@nwpu.edu.cn

[2] Department of Mathematics, Abdul Wali Khan University, Mardan 32300, Pakistan; zahir1987@yahoo.com

* Correspondence: abdulsamadkhan17@mail.nwpu.edu.cn

Received: 5 May 2019; Accepted: 3 June 2019; Published: 12 June 2019

Abstract: The current article discussed the heat transfer and thermal radioactive of the thin liquid flow of Sisko fluid on unsteady stretching sheet with constant magnetic field (MHD). Here the thin liquid fluid flow is assumed in two dimensions. The governing time-dependent equations of Sisko fluid are modeled and reduced to Ordinary differential equations (ODEs) by use of Similarity transformation with unsteadiness non-dimensionless parameter St. To solve the model problem, we used analytical and numerical techniques. The convergence of the problem has been shown numerically and graphically using Homotopy Analysis Method (HAM). The obtained numerical result shows that the HAM estimates of the structures is closed with this result. The Comparison of these two methods (HAM and numerical) has been shown graphically and numerically. The impact of the thermal radiation Rd and unsteadiness parameter St over thin liquid flow is discovered analytically. Moreover, to know the physical representation of the embedded parameters, like β, magnetic parameter M, stretching parameter ξ, and Sisko fluid parameters ε have been plotted graphically and discussed.

Keywords: Sisko fluid; unsteady stretching sheet; thin films; MHD; HAM and numerical method

---

1. Introduction

Recently in few years it has been scrutinized that the analysis of the thin film flow has pointedly contributed in different areas like industries, engineering, and technology, etc. All problems related to thin film flow have varied applications in different fields. Some common usage of thin film associated with daily life are wire and fiber coating, extrusion of polymer and metal from die, crystal growing, plastic sheets drawing, plastic foam processing, manufacturing of plastic fluid, artificial fibers, and fluidization of reactor. Thin polymer films have abundant applications in engineering and technology. Several trade and biomedical sectors are associated with caring and functional coatings, non-fouling bio surfaces, advanced membranes, biocompatibility of medical implants, microfluidics, separations, sensors and devices, and many more. In the light of the above applications, this issue brings the attention of the researchers to improve the development of such type of study. At the initial stage the thin film flow problems discuss for viscous fluid flow and with the passage of time it turned to some Non-Newtonian Fluids. Crane [1] discussed at first time the viscid fluid motion in linear extended plate. Dandapat [2] has discussed the flow of the heat transmission investigation of the viscoelastic fluids over an extended surface. Wang [3] investigate time depending flow of a finite thin layer fluid upon a stretching plate. Ushah and Sridharan [4] used horizontal sheet for the same said problem and prolonged it to the thin partial with the analysis of heat transfer. Liu and Andersson [5] discuss heat transmission investigation of the thin partials motion upon an unsteady extending piece. Aziz et al. [6] investigate the problem related to heat transfer of thin layer flow with thermal radioactivity. Tawade et al. [7] discussed the same problem with external magnetic field.

Furthermore, it is observed that the thin layer flow has vital roles in different fields of science. Andersson [8] is considered to be a pioneer who used the Power law model and investigated the liquid film flow viscid fluids flow upon a time dependent extending piece. Waris et al. [9] have investigated nanofluid flow with variable viscosity and thermal radioactive pass a time dependent and extending Sheet. Anderssona et al. [10] discussed the Heat transfer investigation in liquid film upon time dependent stretching plate. Chen [11,12] examined the same problem using the Power-law model. Wang et al. [13] used HAM Method to discuss the problem as examined by Chen. Saeed et al. [14] recently investigated the thin layer flow of casson Nano fluid with thermal radioactivity. The disk they took was rotating and the flow was three-dimensional. Shah et al. [15] discussed the same problem using Horizontal rotating disk. Khan et al. [16,17] studied the heat transfer investigation of the inclined magnetic field with Graphene Nanoparticles. Ullah et al. [18] studied the Brownian Motion and thermophoresis properties of the nanofluids thin layer flow of the Reiner Philippoff fluid upon an unstable stretching sheet. Shah et al. [19] investigate the thin film flow of the Williamson fluid upon an unsteady stretching surface.

Non-Newtonian fluids have abundant uses in the field of energy and technology. Plastic, food products, wall paint, greases, lubricant oil, drilling mud, etc are some common examples. Sisko fluid is also very important non-Newtonian fluids. At a very low shear rate, the Sisko flow has the same behavior as the Power-Law fluid. This property was used experimentally to fit the data of the flow of lubricating greases and also to model the flow of the whole human saliva [20]. Munir et al. studied Sisko fluid with mixed convection heat transfer. Siddiqui et al. [21,22] studied the thin film flow of Sisko fluid. Khan et al. [23] investigated the flow of boundary layer of the Sisko fluid upon a stretching surface. Molati et al. [24] discussed the MHD problem related to sisko fluid. Malik et al. [25] studied the sisko fluid with heat transfer and using convective boundary conditions.

In the past few years, the study of heat transfer and radiative flow through the stretch sheet has taken the attention from many scientists because of its large applications in engineering and industrial processes [26–31]. Rubber production, colloidal suspension, production of glass socks, metal spinning and drawing of plastic film, textile and paper production, use of geothermal energy, food processing, plasma studies, and aerodynamics are some practical examples of such flows. Radiation is often encountered in frequent engineering problems. Keeping in view its applications, Sheikholeslami et al [32–35] presented the application of radioactive nanofluid flow in different geometries. Recently some researchers studied the problems related to heat transfer. Poom et al. [36] examined the casson nanofluid with MHD radiative flow and heat source using rotating channel. Khan et al. [37] discussed the similar HMD problem as above with time dependent and thin film flow of the three fluid models Oldroyed-B, Maxwell, and Jeffry. Ishaq et al. [38] discussed the MHD effect of unsteady porous stretching surface taking Nanofluid film flow of Eyring Powell fluid. Muhammad et al. [39] discussed the MHD rotating flow upon a stretching surface with radioactively Heat absorption. Hsiao [40–43] discussed non-Newtonian fluid flow with MHD.

The purpose of this manuscript is to model and analyze thin film flow of Sisko fluid on time dependent stretching surface in the presence of the constant magnetic field (MHD). Here the thin liquid fluid flow is assumed in two dimensions. The governing time-dependent equations of Sisko fluid are modelled and reduced to ODEs by use of Similarity transformation with unsteadiness non-dimensionless parameter *St*. To solve the model problem, we used Homotopy Analysis Method-(HAM) which is one of the strongest and most time-saving methods. In addition, the heat transfer rates of thermal radiation are studied and analyzed. Shang [44] used another strong technique as Lie Algebra for solution of such type problem.

2. Basic Equations

The stated governing equation and heat equation are given below [23,24]:

$$divV = 0, \tag{1}$$

$$\rho \frac{dV}{dt} = -\nabla p + divS \tag{2}$$

For Sisko fluid S is given as [25–28]

$$\vec{S} = \left[a + b \left| \sqrt{\frac{1}{2} tr\binom{2}{1}} \right|^{n-1} \right]_1 \tag{3}$$

where

$$A_1 = (gradV) + (gradV)^T \tag{4}$$

wherever a, b, n are the material constants which are distinct for dissimilar fluids. If we take a = 0, b = 1 and n = 0 in the Sisko fluid model then we obtained the Power-law fluid model. If we take a = 1, b = 0 and n = 1 in the Sisko fluid model then we obtained the stress–strain relationship of Newtonian fluid. Because of two-dimensional fluid flow the velocity and the stress profile are presumed as

$$\vec{V} = \left[\vec{u}(x,y), \vec{v}(x,y), 0 \right], \ \vec{S} = \vec{S}(x,y), T = T(x,y) \tag{5}$$

where $\vec{u}\,\&\,\vec{v}$ are representing velocity components.

Inserting Equation (5) into Equations (1) and (2), the momentum and continuity equations reduce to the form as:

$$\vec{u}_x + \vec{v}_y = 0 \tag{6}$$

$$\rho\left(\vec{u}_t + \vec{u}\vec{u}_x + \vec{v}\vec{u}_y\right) = -p_{1_x} + a\left(\vec{u}_{xx} + \vec{u}_{yy}\right) + 2b\frac{\partial}{\partial x}\left[\vec{u}_x\left|4\left(\vec{u}_x\right)^2 + \left(\vec{u}_y + \vec{v}_x\right)^2\right|^{\frac{n-1}{2}}\right]$$
$$+ b\frac{\partial}{\partial y}\left[\left(\vec{u}_y + \vec{v}_x\right)\left|4\left(\vec{u}_x\right)^2 + \left(\vec{u}_y + \vec{v}_x\right)^2\right|^{\frac{n-1}{2}}\right] \tag{7}$$

$$\rho\left(\vec{v}_t + \vec{u}\vec{v}_x + \vec{v}\vec{v}_y\right) = -p_{1_x} + a\left(\vec{v}_{xx} + \vec{v}_{yy}\right) + b\frac{\partial}{\partial x}\left[\left(\vec{u}_y + \vec{v}_x\right)\left|4\left(\vec{u}_x\right)^2 + \left(\vec{u}_y + \vec{v}_x\right)^2\right|^{\frac{n-1}{2}}\right] +$$
$$2b\frac{\partial}{\partial y}\left[\vec{v}_y\left|4\left(\vec{u}_x\right)^2 + \left(\vec{u}_y + \vec{v}_x\right)^2\right|^{\frac{n-1}{2}}\right] \tag{8}$$

If $a = 0$, then above equations become the power law fluid and when $h = 0$, then it reduced Newtonian fluid. Introduce the dimensionless variable as:

$$u = \frac{\vec{u}}{\vec{U}}, v = \frac{\vec{v}}{\vec{U}}, t = \frac{t}{\vec{U}}, x = \frac{x}{L}, y = \frac{y}{L} \text{ and } p = \frac{p}{\rho s \vec{U}} \tag{9}$$

Equations (6) and (7) are written as

$$\frac{\partial v}{\partial t} + u\frac{\partial v}{\partial x} + v\frac{\partial v}{\partial y} = -\frac{\partial p_1}{\partial y} + \varepsilon_1\left(\frac{\partial^2 v}{\partial x^2} + \frac{\partial^2 v}{\partial y^2}\right) +$$
$$\varepsilon_2\frac{\partial}{\partial x}\left[\left(\frac{\partial u}{\partial y} + \frac{\partial v}{\partial x}\right)\left|4\left(\frac{\partial u}{\partial x}\right)^2 + \left(\frac{\partial u}{\partial y} + \frac{\partial v}{\partial x}\right)^2\right|^{\frac{n-1}{2}}\right]$$
$$+ 2\varepsilon_2\frac{\partial}{\partial y}\left[\frac{\partial v}{\partial y}\left|4\left(\frac{\partial u}{\partial x}\right)^2 + \left(\frac{\partial u}{\partial y} + \frac{\partial v}{\partial x}\right)^2\right|^{\frac{n-1}{2}}\right], \tag{10}$$

The dimensionless parameter ε_1 and ε_2 are defined as

$$\varepsilon_1 = \frac{a/\rho}{L\vec{U}} \text{ and } \varepsilon_2 = \frac{b/\rho}{L\vec{U}}\left(\frac{\vec{U}}{L}\right)^{n-1} \tag{11}$$

The above equation after using the boundary layer approximations become as

$$\rho\left(u_t + uu_x + vu_y\right) = -p_{1_x} + au_{yy} + b\frac{\partial}{\partial y}\left(\left|u_y\right|^{n-1}u_y\right) \tag{12}$$

$$0 = -p_{1_y} \tag{13}$$

3. Mathematical Formulation of the Problem

Consider a time depending and electric conducting the flow of thin layer of the Sisko fluid with impact of thermal radiations during spreading surface where x-axis is in parallel to the slot where y-axis shown in the figure is orthogonal to the surface as given below in Figure 1.

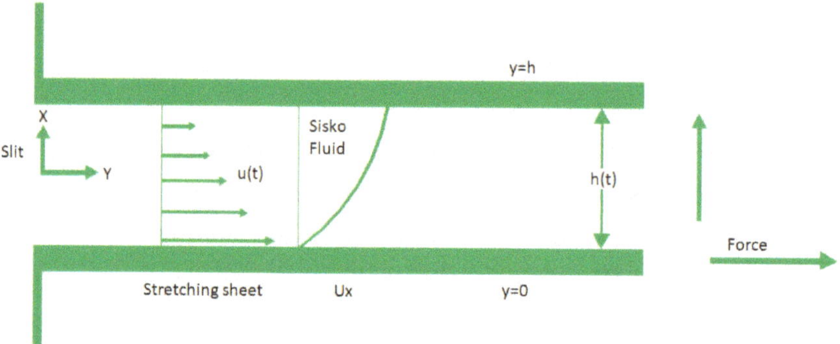

Figure 1. Physical Sketch of the model problem.

It is stretching flow surface also the origin is immovable because equivalent and opposed forces are acting along the X-axis. The x-axis is taking with stress velocity along the spreading surface.

$$U(x,t) = \frac{cx}{1 - bt} \tag{14}$$

In which c and b are constants. The y-axis is making a right angle to it. The term $\frac{cx^2}{v(1-bt)}$ is local Reynolds number, the surface velocity $U(x,t)$. The heat and mass transfer simultaneously here is defined as $T_s(x,t) = T_0 - T_{ref}\ldots\left[\frac{cx^2}{2v}\right](1 - bt)^{\frac{1}{2}}$, which is surface temperature. Here T_0 is temperature at the slit, T_{ref} is reference temperature such that $0 \leq T_{ref} \leq T_0$. also $T_s(x,t)$ defines temperature of the sheet, obtain from the T_0 at the slit in $0 \leq c \leq 1$.

First the slit is static with center and after an exterior power is implied to bounce the slit in the positive x -axis at the rate $\frac{c}{1-bt}$, where $c \in [0,1]$. An exterior transmission magnetic force is given normally to the extending sheet which is presumed to be variable type and selected as

$$B(t) = B_0(1 - bt)^{-0.5} \tag{15}$$

The fluid flow is assumed unsteady laminar and incompressible. The elementary boundary governing equations after using assumption are reduced as

$$u_x + v_y = 0 \tag{16}$$

$$u_t + uu_x + vu_y = \frac{a}{\rho}u_{yy} - \frac{b}{\rho}\frac{\partial}{\partial y}(-u_y)^n - \sigma \vec{B}^2(t)u \tag{17}$$

$$(T_t) + u(T_x) + v\left(T_y\right) = \frac{k}{\rho C_p}\left(T_{yy}\right) - \frac{1}{\rho C_p}\left(q_{r_y}\right) \tag{18}$$

Here q_r approximation of Rosseland, where q_r is define as [29–32],

$$q_{r_y} = -\left(\frac{4\sigma^*}{3k^*}\right)T_y^4 \tag{19}$$

Here T represents the temperature fields, σ^* is the Stefan–Boltzmann constant, K^* is the mean absorption coefficient, k is the thermal conductivity of the thin film. Expanding T^4 using Taylor's series about T_∞ as below

$$T^4 = T_\infty^4 + 4T_\infty^3(T - T_\infty) + 6T_\infty^2(T - T_\infty)^2 + \ldots, \tag{20}$$

Neglecting the higher order terms

$$T^4 \cong -3T_\infty^4 + 4T_\infty^3 T, \tag{21}$$

Using Equation (21) in Equation (20) we get the following

$$q_{r_y} = -\frac{16T_\infty^* \sigma^*}{3K^*}T_{yy}^4, \tag{22}$$

Using Equation (22) in Equation (18) we get the following

$$T_t + uT_x + vT_y = \frac{k}{\rho C_p}T_{yy} - \frac{1}{\rho C_p}\left(\frac{16T_\infty^* \sigma^*}{3K^*}T_{yy}^4\right), \tag{23}$$

The accompanying boundary conditions are given by

$$\vec{u} = \vec{U}., \vec{v} = 0, T = T_s, \text{ at } y = 0, \\ \vec{u}_y = T_y = 0 \text{ at } y = h, \tag{24}$$

Similarity Transformations

The dimensionless variable f and similarity variable η for transformation as

$$.f(\eta) = \psi(x,y,t)\left(\frac{vc}{1-bt}\right)^{-\frac{1}{2}}, \eta = \sqrt{\frac{c}{v(1-bt)}}y, h(t) = \sqrt{\frac{v(1-bt)}{c}};$$

$$\theta(\eta) = T_0 - T(x,y,t)\left(\frac{bx^2}{2v(1-at)^{-\frac{3}{2}}}\left(T_{ref}\right)\right)^{-1} \tag{25}$$

Here $\psi(x..., y..., t...)$ indicate the stream function which identically satisfying Equation (16), $h(t)$ identifies the thin film thickness. The velocity components in term of stream function are obtained as

$$u = \psi_y = \left(\frac{cx}{1-bt}\right) f'(\eta), v = -\psi_x = -\left(\frac{vc}{1-bt}\right)^{\frac{1}{2}} f(\eta),$$
(26)

Inserting the similarity transformation Equation (25) into Equations (16)–(18) and Equation (23) fulfills the continuity Equation (16)

$$\varepsilon f''' + f f'' + n\xi(-f'')^{n-1} f''' - (f')^2 - St\left(\frac{1}{2}\eta f'' + f'\right) - Mf' = 0...,$$
(27)

$$(1 + Rd)\theta'' - \Pr\left(\frac{S}{2}(3\theta + \eta\theta') + 2f'\theta - \theta'f\right) = 0,$$
(28)

The boundary constrains of the problem are:

$$
\begin{aligned}
f'(0) &= 1,\ f(0) = 0, \theta(0) = 1,\\
f''(\beta) &= 0, \theta'(\beta) = 0,\\
f(\beta) &= \frac{S\beta}{2}
\end{aligned}
$$
(29)

The dimensionless film thickness $\beta = \sqrt{\frac{b}{v(1-at)}} h(t)$ which gives

$$h_t = -\frac{\alpha\beta}{2}\sqrt{\frac{v}{b(1-\alpha t)}}$$
(30)

The physical constraints after generalization are obtained as, $St = \frac{\beta}{c}$ is the non-dimensional measure of unsteadiness, $\varepsilon = \frac{a}{\rho v}$ is a Sisko fluid parameter, $\xi = \frac{b}{\rho v}\left(\left(\frac{cx}{(1-bt)\sqrt{v}}\right)^{\frac{3}{2}}\right)^{n-1}$ is stretching parameter, and $M = \frac{\sigma_f B_0^2}{b\rho_f}$ represents the magnetic, $\Pr = \frac{\rho v c_p}{k} = \frac{\mu c_p}{k}$ is the Prandtl number and $Rd = \frac{4\sigma T_s^3}{kk^*}$ represent the radiation parameter.

The Skin friction is defined as

$$C_f = \frac{(S_{xy})_{y=0}}{\rho U_w^2},$$
(31)

where

$$S_{xy} = \mu_0\left(a + b|u_y|^n u_y\right) y = 0$$
(32)

$$C_f \sqrt{Re_x} = \varepsilon f''(0) - (-f''(0))^m$$
(33)

where R_{e_x} is known as the local Reynolds number defined as $Re_x = \frac{U_w x}{v}$. The Nusselt number is defined as $u = \frac{\delta Q_w}{\hat{k}(T-T_\delta)}$, in, which Q_w is the heat flux, where $Q_w = -\hat{k}\left(\frac{\partial T}{\partial y}\right)_{\eta=0}$. After the dimensionalization the u is gotten the below as

$$u = -\left(1 + \frac{4}{3}Rd\right)\Theta'(0),$$
(34)

4. Application of Homotopy Analysis Method

In this section HAM is applied to Equations (27)–(29) to get an approximate analytical solution of MHD Sisko fluid flow over unsteady sheet in a following way:

$$f_0(\eta) = \eta, \theta_0(\eta) = 1,$$
(35)

The linear operators are

$$L_f(f) \;=\; \frac{d^3 f}{d\eta^3}, \; L_\theta(\theta) \;=\; \frac{d^2\theta}{d\eta^2} \tag{36}$$

The above differential operators' contents are shown below as

$$\begin{aligned}
L_f\!\left(\psi_1 + \psi_2\eta + \psi_3\eta^2\right) \;=\; 0,\\
L_\theta(\psi_4 + \psi_5\eta) \;-\; 0
\end{aligned} \tag{37}$$

where $\sum_{i=1}^{5}\psi_i,\, i = 1,2,3\ldots$ are considered as arbitrary constant.

4.1. Zeroth rder Deformation Problem

Expressing $\in [01]$ as an embedding parameter with associate parameters $\hbar_f$, and $\hbar_\theta$ where $\hbar \neq 0$. Then in case of zero order distribution the problem will be in the following form:

$$(1-)L_f\!\left(\hat{f}(\eta,) - f_0(\eta)\right) \;=\; h_f N_f\!\left(\hat{f}(\eta,)\right), \tag{38}$$

$$(1-)L_\theta\!\left(\hat{\theta}(\eta,) - \theta_0(\eta)\right) \;=\; \hbar_\theta N_\theta\!\left(\hat{f}(\eta,), \hat{\theta}(\eta,)\right),$$

The subjected boundary conditions are obtained as

$$\begin{aligned}
f(\eta;P)\big|_{\eta=0} \;=\; 0, \; \frac{\partial f(\eta;P)}{\partial\eta}\Big|_{\eta=0\ldots} \;=\; 1, \; \frac{\partial^2 f(\eta;P)}{\partial\eta^2}\Big|_{\eta=\beta} \;=\; 0,\\
\theta(\eta;P)\big|_{\eta=0} \;=\; 1, \; \frac{\partial\theta(\eta;P)}{\partial\eta}\Big|_{\eta=\beta} \;=\; 0,
\end{aligned} \tag{39}$$

The resultant nonlinear operators have been mentioned as:

$$\begin{aligned}
N_f\!\left(\hat{f}(\eta;)\right) \;=\; \varepsilon f_{\eta\eta\eta} + n\xi\left(-f_{\eta\eta}\right)^{n-1} f_{\eta\eta\eta} + f_{\eta\eta}\hat{f}\\
-\left(\hat{f}_\eta\right)^2 - St\!\left(\hat{f}_\eta + \tfrac{\eta}{2}\hat{f}_{\eta\eta}\right) - M\hat{f}_\eta,
\end{aligned} \tag{40}$$

$$N_\theta\ldots[f(\eta;),\theta(\eta;)] \;=\; (1+Rd)\theta_{\eta\eta} - \Pr\!\left(\frac{S}{2}\left(3\theta + \eta\theta_\eta\right) + 2f_\eta\theta - \theta_\eta f\right) \tag{41}$$

Expanding $\hat{f}(\eta;),\, \hat{\theta}(\eta;)$ in term of with use of Taylor's series expansion we get:

$$\begin{aligned}
f_{\ldots}(\eta;P) \;=\; f_0(\eta) + \sum_{i=1}^{\infty} f_i(\eta)P^i,\\
\theta(\eta;P) \;=\; \theta_0(\eta) + \sum_{l=1}^{\infty} \theta_i(\eta)P^i.
\end{aligned} \tag{42}$$

where

$$f_i(\eta) \;=\; \frac{1}{i!}\frac{\partial f(\eta;P)}{\partial\eta}\Big|_{P=0}, \; \theta_i(\eta) \;=\; \frac{1}{i!}\frac{\partial\theta(\eta;P)}{\partial\eta}\Big|_{P=0} \tag{43}$$

Here $\hbar_f$, $\hbar_\theta$ are selected in a way that the Series (43) converges at $P = 1$, switching $P = 1$ in (43), we obtain:

$$\begin{aligned}
f(\eta,) \;-\; f_0(\eta) + \sum_{i=1}^{\infty} f_i(\eta),\\
\theta(\eta,) \;-\; \theta_0(\eta) + \sum_{i=1}^{\infty} \theta_i(\eta),
\end{aligned} \tag{44}$$

4.2. ith-Order Deformation Problem

Differentiating zeroth order equations i^{th} time we obtained the ith order deformation equations with respect to . Dividing by $i!$ and then inserting $= 0$, so i^{th} order deformation equations

$$
\begin{aligned}
L_f(f_i(\eta) - \xi_i f_{i-1}(\eta)) &= h_f \mathfrak{R}_i^f(\eta), \\
L_\theta(\theta_i(\eta) - \xi_i \theta_{i-1}(\eta)) &= h_\theta \mathfrak{R}_i^\theta(\eta).
\end{aligned}
\tag{45}
$$

The resultant boundary conditions are:

$$
\begin{aligned}
f_i(0) &= f_i'(0) = f_i''(\beta) = 0, \\
\theta_i(0) &= \theta_i(\beta) = 0.
\end{aligned}
\tag{46}
$$

$$
\mathfrak{R}_i^f(\eta) = \varepsilon f_{i-1}''' + n\xi \sum_{j=0}^{i-1}\left((-f'')^{n-1}{}_{i-1-j}f_j'''\right) - \sum_{j=0}^{i-1\Sigma} f_{i-1-j}f' \\
\sum_{j=0}^{i-1\Sigma_{i-1}'} f_{i-1-j}'f_j' - St\left(f_{i-1}' + \tfrac{\eta}{2}f'_{i-1}()\right)
\tag{47}
$$

$$
R_i^\theta(\eta) = (1+Rd)\theta_{i-1}'' - Pr\left[\left(\frac{S}{2}\left(3\theta_{i-1} + \eta\theta_{i-1}'\right)\right) - \sum_{j=0}^{i-1} f_{i-1-j}\theta_j' + 2\sum_{j=0}^{i-1} f_{i-1-j}'\theta_j\right].
\tag{48}
$$

where

$$
\xi_i = \begin{cases} 1, & \text{if } P > 1 \\ 0, & \text{if } P \leq 1 \end{cases}
\tag{49}
$$

4.3. Convergence of Solution

After using the HAM method to calculate these solutions of the modelled function as velocity and temperature, these parameters h_f, h_θ are seen. The responsibility of the computed parameters is to regulate of convergence of the series results. In the conceivable region of h, h -curves of $f''(0)$ and $\theta'(0)$ for 12th order approximation are plotted in Figure 2 for different values of numbers. The h-curves consecutively display the valid area. Table 1 values shows the numerical results of HAM solutions at dissimilar approximation. Its shows that the HAM method is fast convergent.

Table 1. The Homotopy Analysis Method (HAM) convergence table up to 25th order approximations when $\varepsilon = 0.5, \beta = 0.5, \xi = St = M = 0.1$.

Approximation	$f''(0)n = 0$	$f''(0)n = 1$	$f''(0)n = 2$	$f''(0)n = 3$	$\Theta'(0)n = 2$
1	0.300000	0.300000	0.300000	0.300000	−0.24761
5	0.489836	0.454647	0.504214	0.486329	−0.214609
10	0.496178	0.457869	0.519242	1.08915	−0.219032
15	0.496308	0.457908	0.520367	0.48611	−0.218519
20	0.496311	0.457908	0.520488	0.485834	−0.218509
25	0.496311	0.457908	0.520488	0.485834	−0.218509

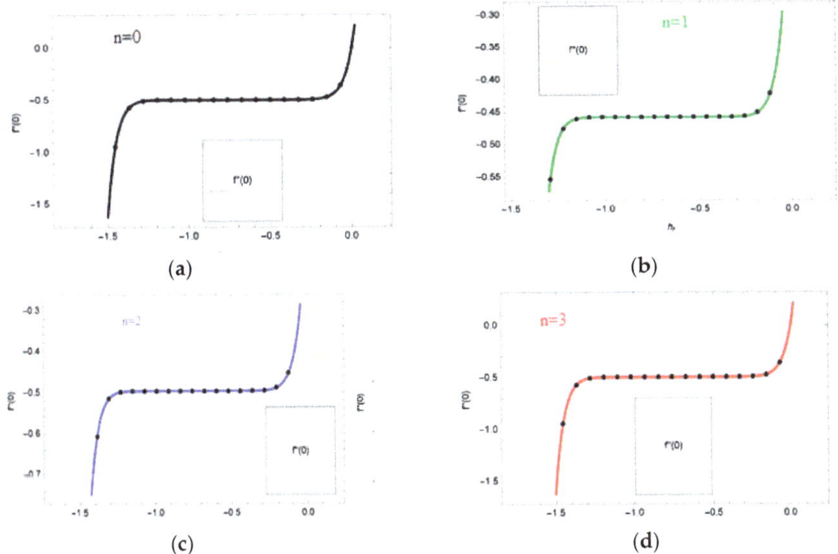

Figure 2. The h-curve graphs for velocity profile for $n = 0, 1, 2, 3$ where $\beta = 0.5, \xi = St = M = 0.1, \varepsilon = 1$.

5. Results and Discussion

The current work investigates the MHD and radiative flow of Sisko thin film flow having unsteady stretching sheet. The purpose of the subsection is to inspect the physical outcomes of dissimilar implanting on the velocity distributions $f(\eta)$ and temperature distribution $\Theta(\eta)$ which are described in Figures 3–12. Figure 3a–d shows the influence of unsteady constraint St on the velocity profile for dissimilar values of power index $n = 0, 1, 2$ and 3. Increasing St increase the velocity field $f(\eta)$. It is clear that varying power indices x having similar response to the time dependent parameter, that is the increase value of the power index rise the velocity distribution. The impact of the unsteadiness parameter St on the heat profile $\theta(\eta)$ is shown in Figure 4. It is observed that $\theta(\eta)$ directly proportional to St. augmented St rises the temperature, which in turn rises the kinetic energy of the fluid, so the fluid motion increased. It is perceived that the effect of St for the different value of power index $n = 0, 1, 2$ and 3 having a similar effect on heat profile $\theta(\eta)$. Figure 5a–d demonstrate the effect of the film thickness β for dissimilar values of power index $n = 0, 1, 2$ and 3. It is perceived that the velocity profile falling down with higher values. Figure 6a–d describe the characteristics for M for changed values of power index $n = 0, 1, 2$ and 3. When M increase on the surface of the sheet during the flow, the flow rate falls, which in results decrease the velocity profiles. This significance of M on velocity field is due to the rise in the M progresses the friction force of the movement, which is called the Lorentz force. It is the fact that fluid velocity reduce in the boundary layer sheet. Figure 7 demonstrates the effect of film thickness β on temperature profile. It is observed that the large values of film thickness β rise the temperature, and actually the higher value of β speed up the molecular motion of the liquids which in turn increases the internal energy and the temperature increase. The effect of stretching parameter ξ for each changed values of power index $n = 0, 1, 2$ and 3 on velocity profile is shown in Figure 8a–c. In case of $n = 1$, it is clear from Figure 8a that velocity profile increase for large value of stretching parameter ξ. When values of power index are varied and the effect of stretching parameter ξ become changed and for $n = 3$ this effect is totally opposite that is the velocity field $f(\eta)$ reduces. For $n = 0$ the stretching parameter ξ becomes zero. The effect of Sisko fluid parameter ε for each changed values of power index $n = 0, 1, 2$ and 3 on velocity profile is shown in Figure 9a–d. The large values of Sisko fluid parameter ε rise the fluid motion, but when the power index goes toward increase then this effect is seen changed and in case of $n = 3$ the velocity field reduces (Figure 9d). The impact of Pr on $\Theta(\eta)$ is shown in Figure 10. Both temperature and concentration distributions vary in reverse form with Pr. When the Pr. number increasing it decrease the temperature distribution, and when the Pr number decreases it increases the temperature distribution. Same effect of Pr on concentration distribution is shown in (32). For increasing values of Pr power index $n = 0, 1, 2$ and 3 thermal radiation increases rapidly. The effect of Rd (thermal radiation parameter) on $\theta(\eta)$ is presented in Figure 11. When the heat transmission coefficient is minor, then the thermal radiation plays an important role in the heat transfer of comprehensive surface. By increasing thermal radiation parameter Rd, its show that it augments the temperature in the fluid film. Due to this rising the rate of cooling is going down. For increasing values of power index $n = 0, 1, 2$ and 3 thermal radiation increases rapidly. The HAM and numerical comparison has been displayed in in Figure 12a–d. Excellent agreement is found between HAM and numerical method.

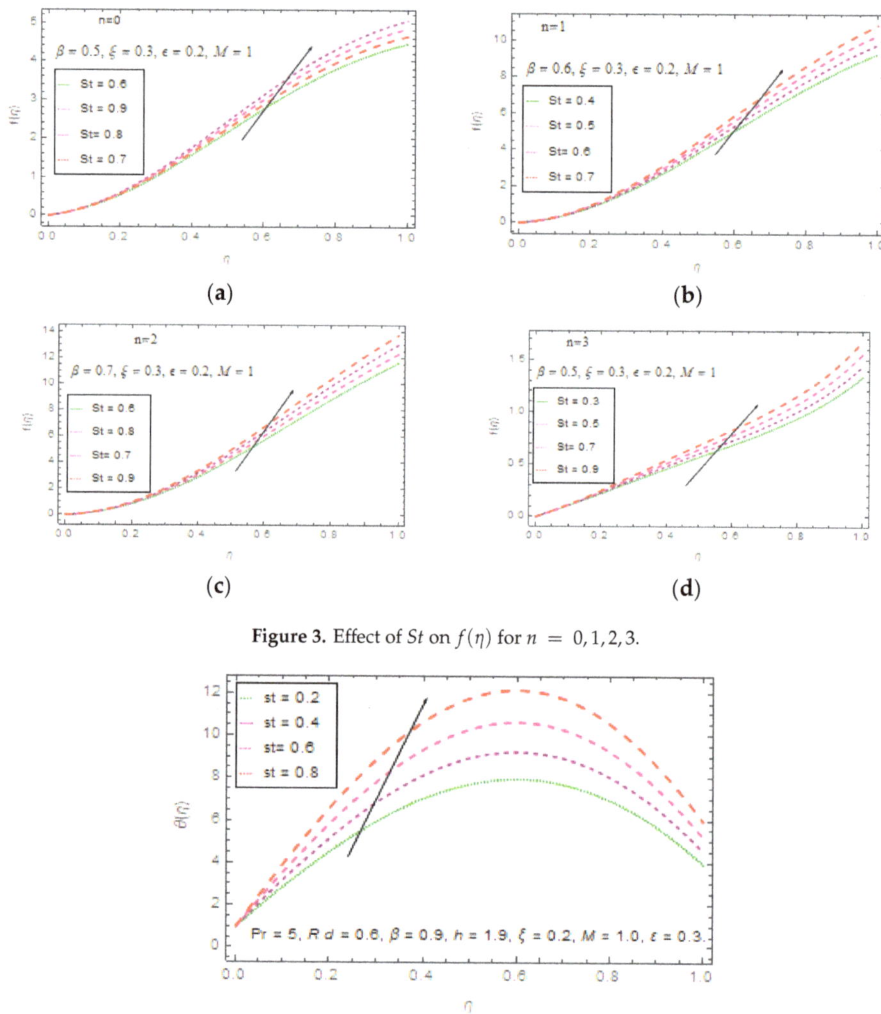

Figure 3. Effect of St on $f(\eta)$ for $n = 0, 1, 2, 3$.

Figure 4. Effect of St on $\theta(\eta)$.

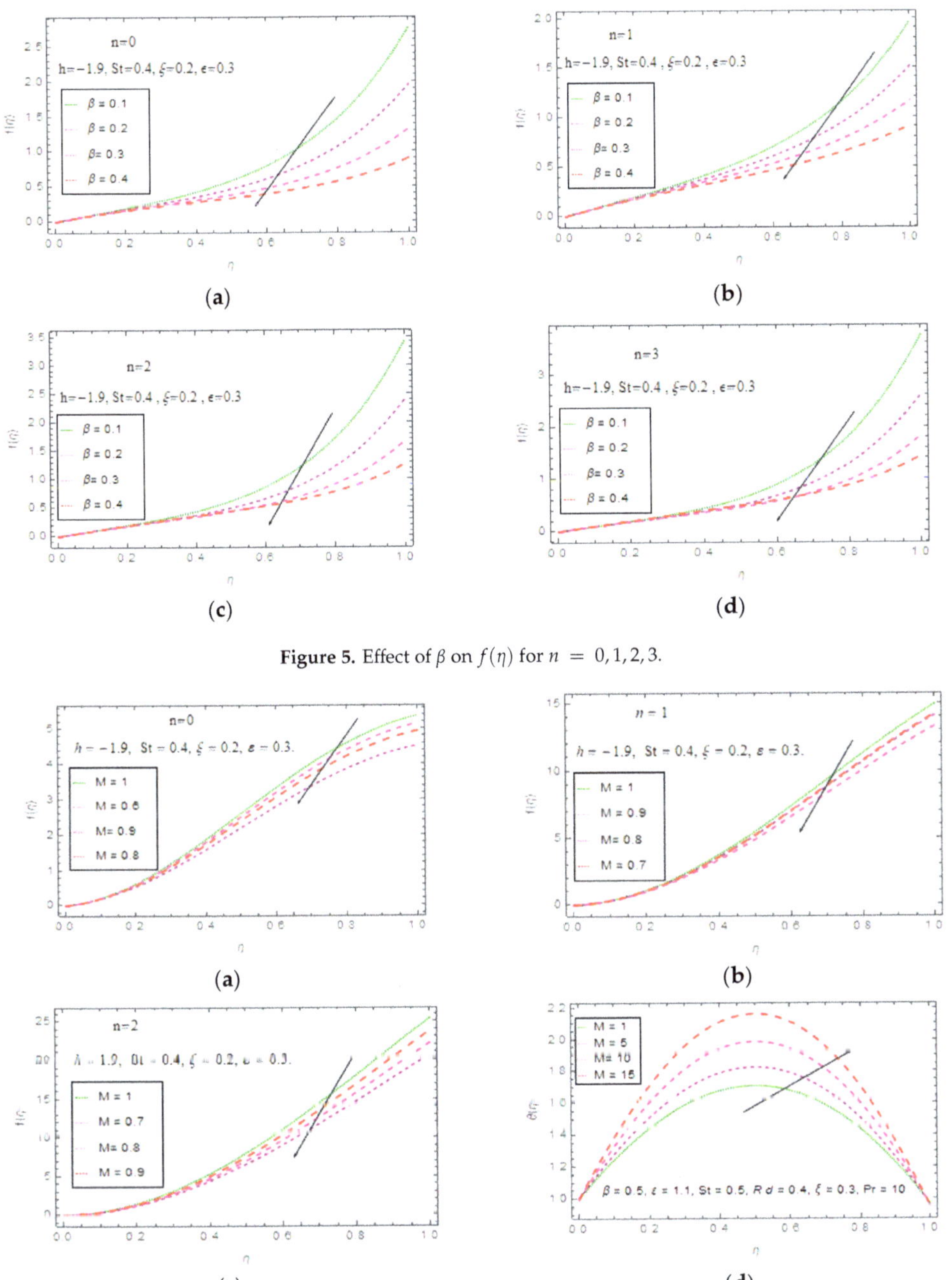

Figure 5. Effect of β on $f(\eta)$ for $n = 0, 1, 2, 3$.

Figure 6. Effect of M on $f(\eta)$ for $n = 0, 1, 2, 3$.

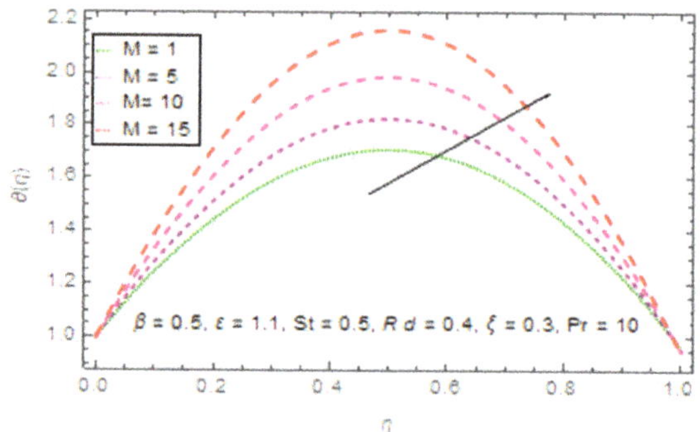

Figure 7. Effect of M on $\theta(\eta)$.

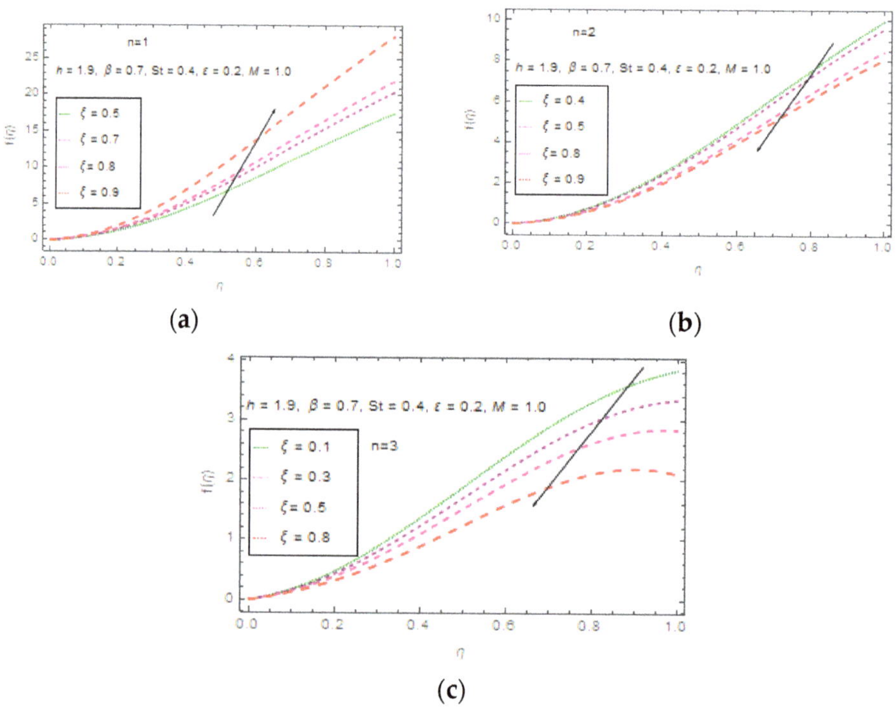

(**a**)

(**b**)

(**c**)

Figure 8. Effect of ξ on $f(\eta)$ for $n = 1, 2, 3$.

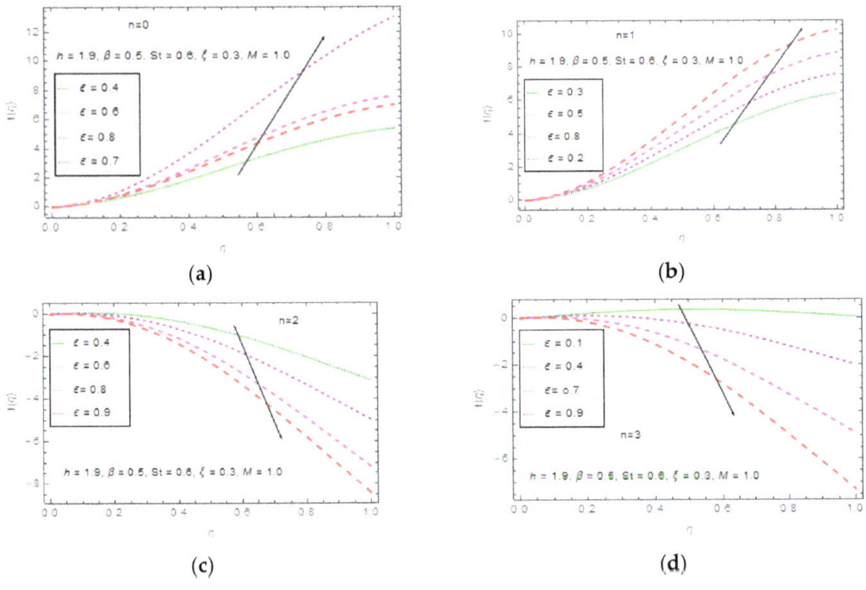

(a)   (b)

(c)   (d)

Figure 9. Effect of ε on $f(\eta)$ for $n = 0, 1, 2, 3$.

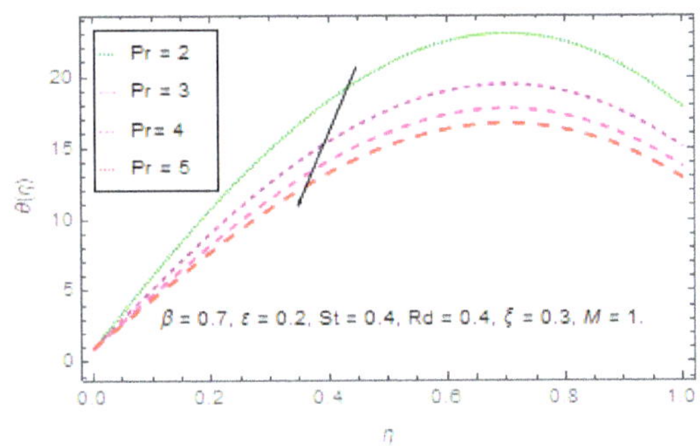

Figure 10. Effect of Pr on $\theta(\eta)$.

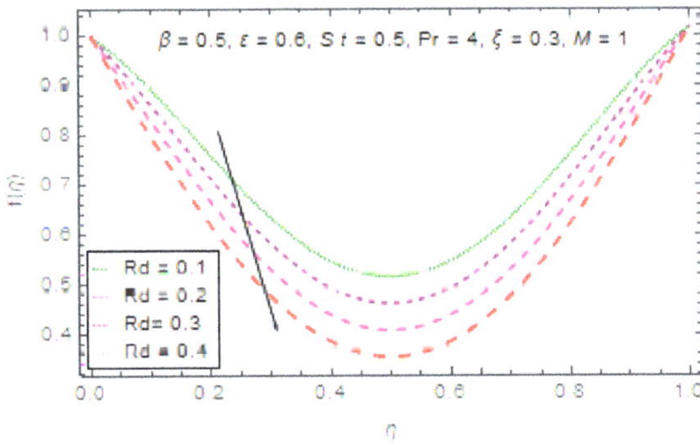

Figure 11. Effect of Rd on $\theta(\eta)$.

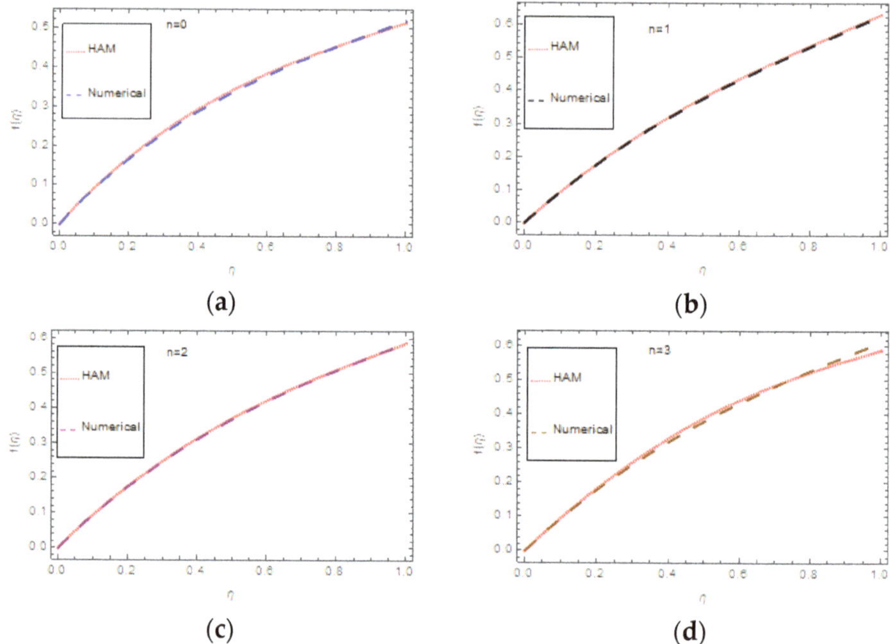

Figure 12. Comparison graphs between Homotopy Analysis Method (HAM) & numerical solution for velocity profiles $f(\eta)$ when $\beta = \varepsilon = 1, h = -0.6, St = M = 0.1, \xi = 0.8$.

Table Discussion

The modelled equations with boundary conditions) are solved analytically as well as numerically. The comparison between analytical and numerical solutions is shown graphically as well as numerically in Tables 1–5 for velocities and temperature. From these tables, an excellent agreement between HAM and Numerical (ND-Solve Techniques) are obtained. The comparison for velocity and temperature profiles for different behavior of fluid between HAM and numerical method are shown in Tables 2–6, which shows an excellent agreement with HAM solution. The numerical values of M, k, β and St on skin friction C_f are given in Table 7. From this table it is clear that increasing values of M, k and β decrease C_f while increasing We increase skin friction. The numerical values of the surface temperature $\theta(\beta)$ for the dissimilar value of M, Rd and St are given in Table 8. It is observed that the increasing values of M, Rd and k increase the surface temperature $\theta(\beta)$, where opposite effect is found for Pr, that is the largest value of Pr reduces the surface temperature $\theta(\beta)$. The gradient of wall temperature $\theta'(0)$ for dissimilar values of embedded parameters Rd, β, Pr, S has been shown in Table 8. It is perceived that larger values of Rd, β and Pr fall the wall temperature and St increase the wall temperature gradient $\Theta'(0)$.

Table 2. The relationship between Homotopy Analysis Method (HAM) and Numerical techniques for $f(\eta)$ in case of $n = 0$, when $\beta = \varepsilon = 1, St = M = 0.1, \xi = 0.8$.

η	HAM Solution $f(\eta)$	Numerical Solution $f(\eta)$	Absolute Error AE
0	$-2.759474 \times 10^{-21}$	0.0000	$-2.759474 \times 10^{-21}$
0.1	0.096682	0.096797	-1.15×10^{-4}
0.2	0.187429	0.187875	-4.46×10^{-4}
0.3	0.273169	0.274142	-9.73×10^{-04}
0.4	0.354701	s0.356381	-1.68×10^{-3}
0.5	0.432726	0.435275	-2.549×10^{-3}
0.6	0.507882	0.511437	-3.555×10^{-3}
0.7	0.580755	0.585433	-4.678×10^{-3}
0.8	0.651911	0.657798	5.887×10^{-3}
0.9	0.721905	0.729058	-7.153×10^{-3}
1	0.791305	0.799754	-8.449×10^{-3}

Table 3. Relationship between Homotopy Analysis Method (HAM) and Numerical techniques for $f(\eta)$ in case of $n = 1$, when $\beta = \varepsilon = 1, St = M = 0.1, \xi = 0.8$.

η	HAM Solution of $f(\eta)$	Numerical Solution $f(\eta)$	Absolute Error AE
0	-4.44031×10^{21}	0.0000	-2.759474×10^{21}
0.1	0.097941	0.096797	-1.15×10^{-4}
0.2	0.192135	0.187875	-4.46×10^{-4}
0.3	0.273169	0.274142	-9.73×10^{-4}
0.4	0.354701	0.356381	-1.68×10^{-3}
0.5	0.432726	0.435275	-2.549×10^{-3}
0.6	0.507882	0.511437	-3.555×10^{-3}
0.7	0.580755	0.585433	-4.678×10^{-3}
0.8	0.651911	0.657798	5.887×10^{-3}
0.9	0.721905	0.729058	-7.153×10^{-3}
1	0.791305	0.799754	-8.449×10^{-3}

Table 4. Relationship between Homotopy Analysis Method (HAM) and ND-Solve for $f(\eta)$ in case of $n = 2$, when $\beta = \varepsilon = 1, St = M = 0.1, \xi = 0.8$.

η	$\theta(\eta)$	Numerical Solution of $\theta(\eta)$	Absolute Error AE
0	1.00000	1.000000	0.000000
0.1	1.04865	1.05256	3.9×10^{-3}
0.2	1.09068	1.09617	5.5×10^{-3}
0.3	1.12653	1.13199	5.5×10^{-3}
0.4	1.15656	1.161	4.4×10^{-3}
0.5	1.18115	1.18401	2.9×10^{-3}
0.6	1.20063	1.20172	1.1×10^{-3}
0.7	1.21532	1.21472	6.0×10^{-4}
0.8	1.2255	1.2235	2.0×10^{-3}
0.9	1.23144	1.22851	2.9×10^{-3}
1	1.23337	1.2301	3.8×10^{-3}

Table 5. Association between Homotopy Analysis Method (HAM) and Numerical techniques for $\theta(\eta)$ in case of $n = 1$, when $\beta = 0.1, \varepsilon = 0.2, St = 0.5, \mathrm{Pr} = 0.5, \xi = 0.3, M = 1$.

η	$\theta(\eta)$	Numerical Solution of $\theta(\eta)$	Absolute Error AE
0	1.0000	1.00000	0.00000
0.1	0.937702	0.9303116	7.4×10^{-3}
0.2	0.886216	0.875477	1.1×10^{-2}
0.3	0.844103	0.832878	1.2×10^{-2}
0.4	0.810135	0.800239	9.9×10^{-3}
0.5	0.783271	0.775633	7.6×10^{-3}
0.6	0.762633	0.757494	5.1×10^{-3}
0.7	0.74749	0.744597	2.9×10^{-3}
0.8	0.737235	0.736031	1.2×10^{-3}
0.9	0.731369	0.73116	2.1×10^{-4}
1	0.72949	0.729589	9.9×10^{-5}

Table 6. Association between Homotopy Analysis Method (HAM) and Numerical techniques for $\theta(\eta)$ in case of $n = 1$, when $\beta = 0.1, \varepsilon = 0.2, St = 0.5, Pr = 0.5, \xi = 0.3, M = 1$.

ξ	$\theta(\eta)$	Numerical Solution $\theta(\eta)$ NN	Error
0	1.0000	1.00000	0.00000
0.1	1.02585	1.02950	0.00365
0.2	1.04897	1.05467	0.00570
0.3	1.06937	1.07579	0.00642
0.4	1.08706	1.09311	0.00605
0.5	1.10202	1.10695	0.00493
0.6	1.11426	1.11759	0.00333
0.7	1.12378	1.12533	0.00155
0.8	1.13058	1.13048	0.00010
0.9	1.13466	1.13334	0.00132
1	1.13602	1.13423	0.00179

Table 7. Coefficient of Skin friction for dissimilar values of M, k, β and S.

M	k	β	S	$f''(0)$ $n=0$	$f''(0)$ $n=1$	$f''(0)$ $n=2$	$f''(0)$ $n=3$
0.1	0.5	1.0	1.5	2.6702	2.6702	3.1102	3.3302
0.5				1.9476	2.6472	2.9988	2.9488
1.0				1.7420	2.4421	2.6728	2.6428
1.5	0.1			2.1299	2.3291	3.4399	4.3399
	0.5			2.3215	2.3223	3.4115	4.3215
	1.0			2.2087	2.2001	3.2087	4.2687
	1.5	0.1		2.6921	2.6957	4.6992	5.6422
		0.5		2.1453	2.1453	4.5556	5.4456
		1.0		2.3986	2.1986	3.8871	4.8911
		1.5	0.1	2.1273	2.1272	3.0173	4.1273
			0.5	2.3592	2.3392	3.3472	4.3572
			1.0	2.5048	2.5048	4.5765	5.1048
			1.5	2.9120	3.0020	5.1982	5.9122

Table 8. Values of $\theta(\beta)$ dissimilar values of M, Pr, Rd and S.

M	Pr	Rd	S	Present Result $\theta(\beta)$	Present Result $\Theta'(0)$
0.0	0.1	1.0	0.1	0.2234	3.6823
1.0				0.4321	3.5412
2.0				0.7123	3.4459
5.0				1.0230	3.8180
1.0	0.01			1.6253	3.4111
	0.1			1.2340	3.2222
	1.0			0.9882	5.3042
	5.0			0.5660	3.2914
	1.0	0.0		0.2209	2.8114
		1.0		0.4320	1.1420
		3.0		0.6741	3.3714
		5.0		0.9922	3.1825
		1.0	0.1	0.0112	2.0114
			0.2	0.2276	2.0005
			0.3	0.5300	3.4114
			0.4	0.7192	3.1127
			0.5	1.2005	2.9914

6. Conclusions

Heat transfer and thermal radiation of the thin liquid flow of Sisko fluid on time dependent stretching surface in the presence of the constant magnetic field (MHD) are investigated. Here the thin liquid fluid flow is assumed in two dimensions. The governing time-dependent equations of Sisko fluid are modeled and reduced to ODEs by use of Similarity transformation with unsteadiness non-dimensionless parameter St. To solve the model problem, we used analytical and numerical techniques. The convergence of the problem has been shown numerically and graphically using Homotopy Analysis Method (HAM).

The main key points are given below:

- In the present investigation we see that due to greater value of magnetic parameter, the velocity distribution of the thin films fluid will be decreasing.
- Increasing thin film thickness decreases the motion of the fluid
- Due to increasing radiation parameters, the Nusselt number increases.
- The increasing values of Pr number, raising the temperature of the surface, also caused the surface temperature to fall down for large values of unsteady parameters.
- The effect of the liquid film flow on the flow of Sisko fluids has been studied graphically and also shown in tables.
- At the end it is also summarized that due to the Lorentz's force the liquid film flow is affected.
- The Sisko fluid parameter increases velocity field.
- The effect of all parameters is shown for dissimilar values of power index $n = 0, 1, 2$ and 3.

Author Contributions: The 1st author "A.S.K." modeled and write the paper (Conceptualization and investigation), the 2nd author "Y.N." review, modify the paper and supervised, the 3rd author "Z.S." solve the problem (Methodology and use software)

Funding: This research was funded by National Natural Science Foundation of China No. [11471262].

Conflicts of Interest: The authors declare no conflict of interest.

References

1. Crane, L.J. Flow past a stretching plate. *Angrew. Math. Phys.* **1970**, *21*, 645–647. [CrossRef]
2. Dandapat, B.S.; Gupta, A.S. Flow and heat transfer in a viscoelastic fluid over a stretching sheet. *Int. J. Nonlinear Mech.* **1989**, *24*, 215–219. [CrossRef]

3. Wang, C.Y. Liquid film on an unsteady stretching surface. *Q. Appl. Math.* **1990**, *48*, 601–610. [CrossRef]
4. Usha, R.; Sridharan, R. On the motion of a liquid film on an unsteady stretching surface. *ASME Fluids Eng.* **1993**, *150*, 43–48. [CrossRef]
5. Liu, I.C.; Andersson, I.H. Heat transfer in a liquid film on an unsteady stretching sheet. *Int. J. Therm. Sci.* **2008**, *47*, 766–772. [CrossRef]
6. Aziz, R.C.; Hashim, I.; Alomari, A.K. Thin film flow and heat transfer on an unsteady stretching sheet with internal heating. *Meccanica* **2011**, *46*, 349–357. [CrossRef]
7. Tawade, L.; Abel, M.; GMetri, P.; Koti, A. Thin filmflow and heat transfer over an unsteady stretching sheet with thermal radiation, internal heating in presence of external magnetic field. *Int. J. Adv. Appl. Math. Mech.* **2016**, *3*, 29–40.
8. Andersson, H.I.; Aarseth, J.B.; Braud, N.; Dandapat, B.C. Flow of a power-law fluid film on an unsteady stretching surface. *J. Non-Newton. Fluid Mech.* **1996**, *62*, 1–8. [CrossRef]
9. Waris, K.; Gul, T.; Idrees, M.; Islam, S.; Khan, I.; Dennis, L.C.C. Thin FilmWilliamson Nanofluid Flow with Varying Viscosity and Thermal Conductivity on a Time-Dependent Stretching Sheet. *Appl. Sci.* **2016**, *6*, 334.
10. Anderssona, H.I.; Aarseth, J.B.; Dandapatb, B.C. Heat transfer in a liquid film on an unsteady stretching. *Int. J. Heat Mass Transf.* **2000**, *43*, 69–74. [CrossRef]
11. Chen, C.H. Heat transfer in a power-law liquid fillm over a unsteady stretching sheet. *Heat Mass Transf.* **2003**, *39*, 791–796. [CrossRef]
12. Chen, C.H. Effect of viscous dissipation on heat transfer in a non-Newtonian liquid film over an unsteady stretching sheet. *J. Non-Newton. Fluid Mech.* **2006**, *135*, 128–135. [CrossRef]
13. Wang, C.; Pop, L. Analysis of the flow of a power-law liquid film on an unsteady stretching surface by means of homotopy analysis method. *J. Non-Newton. Fluid Mech.* **2006**, *138*, 161–172. [CrossRef]
14. Saeed, A.; Shah, Z.; Islam, S.; Jawad, M.; Ullah, A.; Gul, T.; Kumam, P. Three-Dimensional Casson Nanofluid Thin Film Flow over an Inclined Rotating Disk with the Impact of Heat Generation/Consumption and Thermal Radiation. *Coatings* **2019**, *9*, 248. [CrossRef]
15. Shah, Z.; Dawar, A.; Kumam, P.; Khan, W.; Islam, S. Impact of Nonlinear Thermal Radiation on MHD Nanofluid Thin Film Flow over a Horizontally Rotating Disk. *Appl. Sci.* **2019**, *9*, 1533. [CrossRef]
16. Khan, N.S.; Gul, T.; Kumam, P.; Shah, Z.; Islam, S.; Khan, W.; Zuhra, S.; Sohail, A. Influence of Inclined Magnetic Field on Carreau Nanoliquid Thin Film Flow and Heat Transfer with Graphene Nanoparticles. *Energies* **2019**, *12*, 1459. [CrossRef]
17. Khan, N.S.; Zuhra, S.; Shah, Z.; Bonyah, E.; Khan, W.; Islam, S. Slip flow of Eyring-Powell nanoliquid film containing graphene nanoparticles. *AIP Adv.* **2018**, *8*, 115302. [CrossRef]
18. Ullah, A.; Alzahrani, E.O.; Shah, Z.; Ayaz, M.; Islam, S. Nanofluids Thin Film Flow of Reiner-Philippoff Fluid over an Unstable Stretching Surface with Brownian Motion and Thermophoresis Effects. *Coatings* **2019**, *9*, 21. [CrossRef]
19. Shah, Z.; Bonyah, E.; Islam, S.; Khan, W.; Ishaq, M. Radiative MHD thin film flow of Williamson fluid over an unsteady permeable stretching. *Heliyon* **2018**, *4*, e00825. [CrossRef] [PubMed]
20. Isko, A. The flow of lubricating greases. *Ind. Eng. Chem.* **1958**, *50*, 1789–1792.
21. Siddiqui, A.; Ahmed, M.; Ghori, Q. Thin film flow of non-newtonian fluids on a moving belt. *Chaos Solitons Fractals* **2007**, *33*, 1006–1016. [CrossRef]
22. Siddiqui, A.; Ashraf, H.; Walait, A.; Haroon, T. On study of horizontal thin film flow of sisko fluid due to surface tension gradient. *Appl. Math. Mech.* **2015**, *36*, 847–862. [CrossRef]
23. Khan, M.; Shahzad, A. On boundary layer flow of a Sisko fluid over a stretching sheet. *Quaest. Math.* **2012**, *36*, 137–151. [CrossRef]
24. Molati, M.; Hayat, T.; Mahomed, F. Rayleigh problem for a MHD Sisko fluid. *Nonlinear Anal. Real World Appl.* **2009**, *10*, 3428–3434. [CrossRef]
25. Malik, R.; Khan, M.; Munir, A.; Khan, W.A. Flow and Heat Transfer in Sisko Fluid with Convective Boundary Condition. *PLoS ONE* **2014**, *9*, e107989. [CrossRef]
26. Shah, Z.; Islam, S.; Ayaz, H.; Khan, S. Radiative Heat and Mass Transfer Analysis of Micropolar Nanofluid Flow of Casson Fluid between Two Rotating Parallel Plates with Effects of Hall Current. *ASME J. Heat Transf.* **2019**, *141*. [CrossRef]

27. Shah, Z.; Dawar, A.; Islam, S.; Khan, I.; Ching, D.L.C. Darcy-Forchheimer Flow of Radiative Carbon Nanotubes with Microstructure and Inertial Characteristics in the Rotating Frame. *Case Stud. Therm. Eng.* **2018**, *12*, 823–832. [CrossRef]

28. Shah, Z.; Bonyah, E.; Islam, S.; Gul, T. Impact of thermal radiation on electrical mhd rotating flow of carbon nanotubes over a stretching sheet. *AIP Adv.* **2019**, *9*, 015115. [CrossRef]

29. Shah, Z.; Dawar, A.; Islam, S.; Khan, I.; Ching, D.L.C.; Khan, A.Z. Cattaneo-Christov model for Electrical MagnetiteMicropoler Casson Ferrofluid over a stretching/shrinking sheet using effective thermal conductivity model. *Case Stud. Therm. Eng.* **2019**, *13*, 100352. [CrossRef]

30. Khan, A.S.; Nie, Y.; Shah, Z.; Dawar, A.; Khan, W.; Islam, S. Three-Dimensional Nanofluid Flow with Heat and Mass Transfer Analysis over a Linear Stretching Surface with Convective Boundary Conditions. *Appl. Sci.* **2018**, *8*, 2244. [CrossRef]

31. Khan, A.; Shah, Z.; Islam, S.; Khan, S.; Khan, W.; Khan, Z.A. Darcy–Forchheimer flow of micropolar nanofluid between two plates in the rotating frame with non-uniform heat generation/absorption. *Adv. Mech. Eng.* **2018**, *10*, 1–16. [CrossRef]

32. Li, Z.; Sheikholeslami, M.; Shah, Z.; Shafee, A.; Al-Qawasmi, A.-R.; Tlili, I. Time dependent heat transfer in a finned triplex tube during phase changing of nanoparticle enhanced PCM. *Eur. Phys. J. Plus* **2019**, *134*, 173. [CrossRef]

33. Sheikholeslami, M.; Shah, Z.; Shafi, A.; Khan, I.; Itili, I. Uniform magnetic force impact on water based nanofluid thermal behavior in a porous enclosure with ellipse shaped obstacle. *Sci. Rep.* **2019**, *9*. [CrossRef]

34. Sheikholeslami, M.; Shah, Z.; Tassaddiq, A.; Shafee, A.; Khan, I. Application of Electric Field for Augmentation of Ferrofluid Heat Transfer in an Enclosure Including Double Moving Walls. *IEEE Access* **2019**, *7*, 21048–21056. [CrossRef]

35. Kumam, P.; Shah, Z.; Dawar, A.; Ur Rasheed, H.; Islam, S. Entropy Generation in MHD Radiative Flow of CNTs Casson Nanofluid in Rotating Channels with Heat Source/Sink. *Math. Probl. Eng.* **2019**, *2019*, 9158093. [CrossRef]

36. Khan, A.S.; Nie, Y.; Shah, Z. Impact of Thermal Radiation and Heat Source/Sink on MHD Time-Dependent Thin-Film Flow of Oldroyed-B, Maxwell, and Jeffry Fluids over a Stretching Surface. *Processes* **2019**, *7*, 191. [CrossRef]

37. Ishaq, M.; Ali, G.; Shah, S.I.A.; Shah, Z.; Muhammad, S.; Hussain, S.A. Nanofluid Film Flow of Eyring Powell Fluid with Magneto Hydrodynamic Effect on Unsteady Porous Stretching Sheet, Punjab University. *J. Math.* **2019**, *51*, 131–153.

38. Muhammad, S.; Ali, G.; Shah, Z.; Islam, S.; Hussain, S.A. The Rotating Flow of Magneto Hydrodynamic Carbon Nanotubes over a Stretching Sheet with the Impact of Non-Linear Thermal Radiation and Heat Generation/Absorption. *Appl. Sci.* **2018**, *8*, 482. [CrossRef]

39. Hsiao, K.L. To Promote Radiation Electrical MHD Activation Energy Thermal Extrusion Manufacturing System Efficiency by Using Carreau-Nanofluid with Parameters Control Method. *Energy* **2017**, *130*, 486–499. [CrossRef]

40. Hsiao, K.L. Combined Electrical MHD Heat Transfer Thermal Extrusion System Using Maxwell Fluid with Radiative and Viscous Dissipation Effects. *Appl. Therm. Eng.* **2016**, *112*, 1281–1288. [CrossRef]

41. Hsiao, K.L. Micropolar Nanofluid Flow with MHD and Viscous Dissipation Effects Towards a Stretching Sheet with Multimedia Feature. *Int. J. Heat Mass Transf.* **2017**, *112*, 983–990. [CrossRef]

42. Hsiao, K.L. Stagnation Electrical MHD Nanofluid Mixed Convection with Slip Boundary on a Stretching Sheet. *Appl. Therm. Eng.* **2016**, *98*, 850–861. [CrossRef]

43. Shang, Y. A Lie algebra approach to susceptible-infected-susceptible epidemics, Electronic. *J. Differ. Equ.* **2012**, *233*, 1–7

44. Shang, Y. Analytical solution for an in-host viral infection model with time-inhomogeneous rates. *Acta Phys. Pol. B* **2015**, *46*, 1567. [CrossRef]

 *processes*

Article

Viscoelastic MHD Nanofluid Thin Film Flow over an Unsteady Vertical Stretching Sheet with Entropy Generation

Asad Ullah [1,2], **Zahir Shah** [1], **Poom Kumam** [3,4,5,*], **Muhammad Ayaz** [1], **Saeed Islam** [1] and **Muhammad Jameel** [1]

[1] Department of Mathematics, Abdul Wali Khan University, Mardan 23200, Khyber Pakhtunkhwa, Pakistan; asad.ullah@kust.edu.pk (A.U.); Zahir1987@yahoo.com (Z.S.); mayazmath@awkum.edu.pk (M.A.); saeedislam@awkum.edu.pk (S.I.); muhammadjameel198@gmail.com (M.J.)

[2] Institute of Numerical Sciences, Kohat University of Science & Technology, Kohat 26000, Khyber Pakhtunkhwa, Pakistan

[3] KMUTT-Fixed Point Research Laboratory, Room SCL 802 Fixed Point Laboratory, Science Laboratory Building, Department of Mathematics, Faculty of Science, King Mongkut's University of Technology Thonburi (KMUTT), 126 Pracha-Uthit Road, Bang Mod, Thrung Khru, Bangkok 10140, Thailand

[4] KMUTT-Fixed Point Theory and Applications Research Group, Theoretical and Computational Science Center (TaCS), Science Laboratory Building, Faculty of Science, King Mongkut's University of Technology Thonburi (KMUTT), 126 Pracha-Uthit Road, Bang Mod, Thrung Khru, Bangkok 10140, Thailand

[5] Department of Medical Research, China Medical University Hospital, China Medical University, Taichung 40402, Taiwan

* Correspondence: poom.kum@kmutt.ac.th; Tel.: +66-2-4708-994

Received: 3 April 2019; Accepted: 28 April 2019; Published: 6 May 2019

Abstract: The boundary-layer equations for mass and heat energy transfer with entropy generation are analyzed for the two-dimensional viscoelastic second-grade nanofluid thin film flow in the presence of a uniform magnetic field (MHD) over a vertical stretching sheet. Different factors, such as the thermophoresis effect, Brownian motion, and concentration gradients, are considered in the nanofluid model. The basic time-dependent equations of the nanofluid flow are modeled and transformed to the ordinary differential equations system by using similarity variables. Then the reduced system of equations is treated with the Homotopy Analysis Method to achieve the desire goal. The convergence of the method is prescribed by a numerical survey. The results obtained are more efficient than the available results for the boundary-layer equations, which is the beauty of the Homotopy Analysis Method, and shows the consistency, reliability, and accuracy of our obtained results. The effects of various parameters, such as Nusselt number, skin friction, and Sherwood number, on nanoliquid film flow are examined. Tables are displayed for skin friction, Sherwood number, and Nusselt number, which analyze the sheet surface in interaction with the nanofluid flow and other informative characteristics regarding this flow of the nanofluids. The behavior of the local Nusselt number and the entropy generation is examined numerically with the variations in the non-dimensional numbers. These results are shown with the help of graphs and briefly explained in the discussion. An analytical exploration is described for the unsteadiness parameter on the thin film. The larger values of the unsteadiness parameter increase the velocity profile. The nanofluid film velocity shows decline due the increasing values of the magnetic parameter. Moreover, a survey on the physical embedded parameters is given by graphs and discussed in detail.

Keywords: entropy generation; second-grade fluid; nanofluid; liquid films; thin film; time depending stretching surface; magnetic field; HAM

---

1. Introduction

In the last few years, thin film flow problems have received great attention. The history behind such a loyalty and importance is the use of thin film flow in various technological disciplines. Thin film flow problems cannot be categorized and classified in a simple way, because they are rooted in particular to broad areas, such as from the analysis of flow in human lungs to industrial problems involving lubricants. Examining thin film flow of liquids and its uses leads us to an important relationship between structural mechanics and fluid mechanics. The polymers and metal extraction, drawing of elastic sheet, exchanges, foodstuff striating, fluidization of the devices, and constant forming are some common uses and applications of liquid film flow. In view of these practical uses of liquid film flow, further advancement and development is observed to be necessary. For this purpose, a variety of attempts have been made with constructive geometries from time to time by many investigators. One such an important geometry is the expanding sheet, which has received great attention and become a problem of interest for the investigators [1,2].

In the beginning, thin liquid film flow was devoted to fluids with some viscosity. Classifications of these fluids based on viscosity have made the area saturated. With the passage of time, the domain was extended to non-Newtonian fluids. Non-Newtonian nano-liquids are studied and are discussed with variations in internal and external agents. The transfer of heat is investigated for non-Newtonian nanoliquid thin film flow by Sandeep et al. [3]. In stretching sheet problems, the geometry of the problem is important, due to its time dependency as well as the nature of the sheet. Wang [4] investigated the liquid thin film flow past a time-dependent expanding sheet. The same geometry with finite thin liquid is investigated by Usha et al. [5]. They analyzed the flow on an unsteady stretching sheet for finite thin liquids. Liu et al. [6] investigated the thin film flow for heat transfer enhancement through an expanding sheet. Aziz et al. [7] described the flow of thin fluid film with generation of heat over an expanding surface. Tawade et al. [8] examined thin film liquid flow for the transfer of heat in the presence of thermal radiations. They implemented the RK-Fehlberg and Newton-Raphson method to tackle the modeled equations. A briefer survey on heat transfer analysis on liquid film flow past an expanding sheet is presented by Anderssona et al. [9]. The study of expanding sheet problem is not rare in the literature and a brief survey can be found on its applications and other technological advancements in [10–15]. Besides all these, a variety of fluids are investigated under the same geometry. Megahed [16] studied the impact of heat on thin Casson fluid past an unsteady expanding surface by assuming the slip velocity in the presence of viscous dissipation and heat flux. Shah et al. [17] described the flow of Casson nanofluid in rotating parallel plates in the presence of Hall effect. Jawad et al. [18] studied the MHD flow of the nanofluid thin film by considering the Joule heat loss and Navier's partial slip by considering Darcy-Forchheimer model. Interested readers are referred to [19–22] for more brief discussion on rotating systems. Some new modifications are made by Khan et al. [23] and Tahir et al. [24] for the thin film flow of nanofluids.

The proficiency in different applications of nanofluids are due to the enormous features (heat transfer enhancement, cooling etc.). From the practical application point of view, nanofluids are used in powered engines, pharmaceutical procedures, micro-electrons, and hybrid fuel cells. Presently, its major use is in the field of nanotechnologies. In industries, the use of electronic equipment and nanoboards are essential presently. These boards and electronic accessories become hot with the passage of time, due to which its efficiency is normally badly affected. To overcome this situation, nanofluids are used as a coolant to reduce heat [25]. A literature survey shows that air is used as a coolant in many processes. In the enhancement of the performance of microchips, projectors and LED nanotechnology coolants are used [26,27]. Non-Newtonian fluid flow is found in abundance in the literature and deeply depends on the mechanism in which it is used. One such a mechanism is the peristaltic mechanism, which plays a vital role in physiological and industrial processes. In this process, along the wall of the channel sinusoidal waves are propagated. The best examples of such waves in practice are dialysis, hose pumps, and the heating of lungs etc. Further investigation of this study leads researchers to examine MHD flows. MHD analysis of peristaltic flows plays a key role in medicine and bio-engineering. The variations in viscosity for peristaltic flow are examined by Srivastava et al. [28]. The variation of viscosity with temperature is studied by Abbasi et al. [29] for nanofluid flow.

From earlier study, nanofluids have been found to retain dimensions smaller than 100 nm [7,8]. Nanofluids are known to be a mixture of nanoparticles and the general heat transfer fluids, for example oil, ethylene glycol, glycol, and water etc. Nanoparticles can be prepared in laboratories and in industries on a large scale. It can be obtained from metals such as Ag, Al, Au, Cu, and oxides of metals such as Fe_3O_4, CuO, TiO_2, Al_2O_3, nitrides such as AlN, SiN, carbides (SiC) etc. The nanoparticles obtained from these materials are used in very small amounts for the improvement of heat transfer, due their high thermal conductivity. The enhancement of heat transfer due to thermal systems for augmentation are presently become widespread. Abolbashari et al. [30] described nanofluid flow using Buongiorno's model through a time-dependent stretching sheet. Hayat et al. [31] discussed the nanoliquid flow in three dimensions by using the Maxwell model. Malik et al. [32] studied a mixed convective MHD nanofluid flow on a stretching surface by considering Erying-Powell fluid. Nadeem et al. [33] discussed the Maxwell liquid film flow of nanoparticles past a perpendicular stretching surface. Raju et al. [34] examined the non-Newtonian nanoliquid MHD flow through a cone with free heat convection and mass transfer. The heat transfer investigations of the nanofluid flow through plates are performed by Rokni et al. [35]. Numerical investigations of the non-Newtonian nanoliquid flow on the stretching surface are presented by Nadeem et al. [36]. Shehzad et al. [37] examined the nanoliquid MHD flow of Jaffrey fluid in the presence of convective-type boundary constraints. Sheikholeslami et al. [38,39] analyzed the heat effects on nanoliquid flow by applying an external magnetic field. Mahmoodi et al. [40] studied the flow of nanofluid for cooling purposes and discussed the heat sink for the flow field. The impact of thermal radiations and Hall current is recently explored by Shah et al. [41,42] for rotating surfaces. A detailed study on rotating surfaces with stretching sheet performed by Shah et al. can be found in [43–49]. CuO containing nanofluid thin film flow inside a semi-annulus region is examined by Sheikholeslami and Bhatti [50] numerically with constant magnetic field. They analyzed the heat transfer enhancement and found some good results. Tube-in-tube analysis of heat exchanger for $\gamma-\text{ALOOH}$ nanofluid is performed by Monfared et al. [51]. They found both upper and lower boundaries of irreversibility for platelet and spherical shape geometries. A brief and detailed survey on nanofluids of Sheikholeslami with modern applications of dissimilar phenomena with a variety of approaches can be found in [52,53]. Besides the theoretical study, literature is rich with experimental results on nanofluid flow and its use in the heat transfer analysis. A combined effect of the silver and carbon nanotubes by taking water as the base fluid is performed by Munkhbayar [54]. In this study, he used 3% of the nanoparticles by volume as compared with base fluids. The use of these combined nanoparticles enhances the heat transfer up to 14.5% at a very low temperature. The increase in concentration and heat transfer for a hybrid of the copper oxide and titanium nanoparticles is achieved by taking water and ethylene glycol mixture as the base fluid studied by Hemmat et al. [55]. They performed this experiment in a laboratory from (30–60)°C. At the upper bound of the temperature range they found a 41.5% increase in the thermal conductivity. Hybrid base fluid (water-ethylene) and hybrid nanofluids (titanium-MWCNT's) investigation was carried out by Akhgar et al. [56] for the stability of base fluid and enhancement of thermal conductivity. They observed a 38.7% increase in the thermal conductivity of the nanofluids as compared with the base fluid. Similar experimental results for the enhancement of the thermal conductivity of nanofluids can be found in [57,58].

The free existence of non-Newtonian fluids and its use attracted researchers to construct models and further developments due to its application in industry. Most of the organic compounds came under the umbrella of the non-Newtonian fluids. Food products, molten plastic, wall paints, lubricant oils, drilling mud, and molten plastic are some of the widely used examples non-Newtonian fluids. Surveys suggest that to classify the non-Newtonian fluids in terms of behavior, many models have been introduced and developed. Some of them, Williamson fluid, Walter's-B fluid, Casson fluid,

Carreau fluid, etc., are very common in use. Carreau fluid model is also known as the Newtonian generalized model [59]. The significance of the Carreau fluid model in the field of melts, water-based polymers, and suspensions attracted investigators. Considering the effectiveness of this model, many researchers investigated the nature of Carreau fluid by using different geometries. Some surveys related to this model are presented here. Kefayati et al. [60] performed a survey on the thermosolutal forced convective flow over two circular cylinders with magnetic effects by taking the Carreau fluid model. They also analyzed the entropy generation in their study. Olajuwon [61] studied the Carreau liquid flow over a perpendicular permeable surface with magnetic effect. Hayat et al. [62] investigated this model for a free convection flow over a non-stationary surface. The study of nanofluids is not just linked to the fluids model used, but purely depends on the nature of the nanomaterial used. The shape of the nanomaterial used is more important in the study of heat transfer processes. To enhance the heat transfer, and to improve thermal and hydraulic properties, the shape of the nanoparticles used is also important. Alsarraf et al. [63] implemented a two-phase mixture model to investigate the double-pipe flow of boehmite alumina nanofluid. They presented the results in the form of percentages for both spherical and platelet-shaped nanoparticles under high Reynolds numbers. Similarly, an experimental study is presented by Azari et al. [64] for alumina nanofluid flow. They successfully analyzed the two-phased model theoretical results with practically obtained information.

Due to its complicated nature, non-Newtonian fluids have been studied by many researchers, just for the purpose of explicitly or implicitly explaining the strain rate. An important type of non-Newtonian fluids is Sisko fluid, which has great significance in engineering as well as in technology. Stretching surface analysis was made by Munir et al. [65] by using Sisko fluid bidirectional flow. Sisko model is used by Olanrewaju et al. [66], for the unsteady non-convective fluid flow over a flat surface by taking into account heat transfer. Khan et al. [67] studied the effect of heat energy transfer in an annular pipe of the Sisko fluid steady state flow. Khan et al. [68] described the Sisko fluid boundary-layer flow over a stretching surface. Similar investigation on the stretching surface for the laminar flow by using Sisko fluid model is carried out by Patel et al. [69]. Darji et al. [70] examined the natural convective time-independent Sisko fluid flow of boundary-layer type. Analytical solutions of the Sisko fluid thin film flow for the drainage down a vertical belt is presented by Siddiqui et al. [71]. The MHD Sisko fluid numerical study for an annular region flow is carried out by Khan et al. [72]. Sar et al. [73] studied both the Lie group and Sisko fluid boundary-layer equations. Haneef et al. [74] investigated the Maxwell nanofluid flow for magnetic and electric effects on a stretching plate. Moallemi et al. [75] studied the flow of Sisko fluid in the pipe and discussed the exact solution. Dawar et al. [76] investigated the CNT Casson fluid flow with MHD in a rotating channel for heat transfer analysis. Shah et al. [77] implemented Cattaneo-Christov model for the heat transfer analysis of MHD micropoler Casson fluid over a stretching sheet. Khan et al. [78] investigated the Eyring-Powell nanoliquid film slip flow by considering nanoparticles of graphene. Recently, Khan and Pop [79] investigated the thermophoresis effect and Brownian motion of a boundary-layer nanofluid flow past a stretching sheet. For the enhancement of heat transfer, researchers started to use impurities. Osiac [80] discussed the electrical and structural properties of nitrogen. Radwan et al. [81] performed the synthesis classification and applications of polystyrene. Coating applications of thin film flow and flexible coating with conductive fillers and applications of thermoelectric materials are discussed [82–85].

The best possible design conduction in the energy system is always the aim of investigators. For this purpose, the role of entropy cannot be ignored in modeling, and other optimization applications of the energy systems. The roots of entropy are connected to the second law of thermodynamics and its irreversible aspects are laid down by Kelvin and Clausius. The theoretical background of entropy was developed very rapidly and has been extended to the new generation. However, in the heat transfer process/thermal radiations, entropy generation cannot be treated by conventional thermodynamic approaches. That is why researchers are compelled to take an interest in the second law of thermodynamics and to investigate its engineering applications. When there is a heat transfer in the system, there must be the generation of entropy. The generation of entropy is mainly concerned with the irreversibility of thermodynamics. Entropy can be generated from different sources, such as viscous dissipation, mass diffusion, and finite temperature gradients in the transfer of heat. The generation of entropy in thermal engineering has been investigated by Bejan [86,87] from some new aspects. The work already available in the system vanishes due to the generation of entropy. From an engineering point of view, it makes sense to understand the irreversibility and mechanism of the entropy generation in the transfer of heat and other problems in the fluid flow.

The generation of entropy in thermal systems has been discussed by many investigators. Weigand and Birkefeld [88] investigated the laminar flow past a flat plate with entropy generation. Makinde [89] investigated the second law for the hydro-magnetic flow of the boundary layer on a stretching surface along with the heat transfer analysis by using variable viscosity. Makinde reported that with the increasing values of Prandtl number and radiation parameter the entropy generation decreases. Hayat et al. [90] presented Darcy-Forchheimer CNT-based nanomaterial convective flow with heat flux for entropy generation. They analyzed both SWCNTs and MWCNTs for heat transfer enhancement and entropy generation. In another investigation of gravity-driven thin film flow in the direction of the heated inclined plate, Makinde [91] reported that at the liquid-surface, the irreversibility of heat transfer is dominant, while an opposite result is observed at the surface of the plate. A small amount of work is done on the heat transfer analysis, considering the second law of thermodynamics in nanofluids. This is because of the rareness of the nanofluids. Recently, Esmaeilpour and Abdollahzadeh [92] investigated the enhancement of heat for nanofluid free convection flow inside an enclosure with the entropy generation. Dawar et al. [93] presented a semi-analytic solution to the CNT nanofluid flow inside rotating plates. In this investigation, the effect of magnetic field and entropy generation is deliberated numerically. A further brief survey on entropy generation by investigating different models and geometries can be found in [94,95]. Surveys suggest that the stretching sheet problem has been studied by different researchers for different fluids. Entropy generation of viscoelastic fluid for the second-grade fluid is rare and will be discussed in this article for the first time. In science and engineering majority of the mathematical models are very complicated, and it is also impossible to solve these types of problems for exact solution. For this purpose, researchers are widely using numerical and semi-analytic methods for the approximate solution. Numerical techniques are sometimes difficult to perform in an efficient way for some problems. This happens due the high non-linearity of the problem. To overcome this situation, Liao [96,97] investigated the solution of these types of problems by implementing a new technique. The method is termed the Homotopy Analysis Method (HAM), due to the use of homotopy, a topological property. He further discussed the convergence of the new implemented method. A solution is a function of single variable in the form of a series.

The foremost aim of discussing this work is to investigate the flow of second-grade thin film flow in the presence heat transfer and magnetic effect with the generation of entropy in a vertical stretching sheet. To the best of our knowledge, we cannot find new work on nanofluid thin film flow on vertical stretching sheet by considering viscoelastic fluid of second grade. Equations for the generation of entropy and the boundary-layer heat transfer over a vertical stretching sheet are constituted for two-dimensional nanofluid thin film flow with uniform magnetic field (MHD). These equations are the leading-order mathematical equations constituted from the geometry, keeping in view the assumptions in flow field. Different physical aspects such as thermophoresis, concentration gradients, and Brownian motion of the flow are assumed in the nanofluid model. The modeled leading-order system in the form of PDEs (partial differential equations) are further transformed into a system of ODEs (ordinary differential equations), with the help of similarity variables. Similarity variables have the property of non-dimensionalization and transforming the system of PDEs to a single independent variable system also known as ODEs. The reduced system of differential equations is tackled by an analytic approach (HAM). HAM is implemented with initial guesswork, as required for the implementation of the technique, due to its fast convergence. The convergence of the implemented technique is discussed numerically and with graphs. The state variables are plotted under the variation of different physical parameters and discussed in detail. Physical quantities such as Sherwood number, Nusselt number, and skin friction are presented numerically for its significance in the boundary-layer flow. The impact of Brinkman number and Bejan number is discussed by graphs for entropy function. The effect of Reynolds number, Prandlt number, and magnetic parameter is also influential on entropy function, Bejan number, and Brinkman number by graphs.

2. Problem Formulation

Assume an unsteady nanoliquid thin film flow of the second-grade fluid past a stretching sheet. Furthermore, the fluid is assumed to be electrically conducting and the magnetic field effect is also considered inside the flow field. The moving sheet starts its motion from a fixed slit. The arrangement of the geometry is made in the Cartesian system of coordinates in such a way that the plate length is equal to ox, and oy is flat to the surface. The stretching effects are applied in such a manner to the surface of the flow that the two forces are in opposite direction with an equal magnitude along the x-axis, and keeps the center motionless. The stretching sheet and x-axis are taken in such a manner that it is adjacent to each other, and the stress velocity of the sheet is given by [17]:

$$U_w(x,t) = \gamma x (1 - \zeta t)^{-1}, \tag{1}$$

where ζ and γ represents any fix numbers, which are vertical to x-axis. The wall temperature and capacity of the nanoparticles are given by [17,19]:

$$T_w(x,t) = T_r \left(\frac{\gamma x^2}{2\nu_f} \right) (1 - \zeta t)^{-1.5} + T_0 \tag{2}$$

$$C_w(x,t) = C_r \left(\frac{\gamma x^2}{2\nu_f} \right) (1 - \zeta t)^{-1.5} + C_0 \tag{3}$$

where ν_f denotes the fluid kinematic viscosity, T_0 and C_0 denotes the temperature of the slit and volume friction of the nanoparticles, while T_r and C_r represents the reference temperature and reference volume of the nanoparticles, respectively.

The applied magnetic field has the relationship of the form [17]:

$$B(t) = B_0 (1 - \zeta t)^{-\frac{1}{2}}, \tag{4}$$

here B_0 is the magnetic field strength.

The mathematical model that illustrates the second-grade fluid is given in the form [23]:

$$\vec{T} = -p\vec{I} + \vec{S}. \tag{5}$$

The basic equations of the second-grade fluid can be obtained by using $\beta_1 = \beta_2 = \beta_3 = 0$ in the third-grade fluid model *i.e.*

$$\vec{S} = \tau = \mu \vec{A}_{(1)} + \alpha_{(2)} \vec{A}_{(2)} + \alpha_{(2)} \vec{A}_{(2)}^2. \tag{6}$$

Here $p\vec{I}$ is an isotropic stress, $\vec{S}$ is an extra tensor of stress, μ is the coefficient of viscosity, $\alpha_{(1)}$ and $\alpha_{(2)}$ represent the thermal stress moduli, $\vec{A}_{(1)}$ and $\vec{A}_{(2)}$ are the stresses of Rivlin Ericksen tensors and have the mathematical structures $\vec{A}_{(1)} = \nabla \vec{V} + (\nabla \vec{V})^T$ and $\vec{A}_{(2)} - \frac{D}{Dt}(\vec{A}_{(1)}) + \vec{A}_{(1)} \nabla \vec{V} + (\nabla \vec{V})^T \vec{A}_{(1)}$

2.1. Continuity Equation:

The equation of continuity under the assumed assumption (in-compressible fluid) for the modeled geometry takes the vectorial form [74]:

$$\nabla . \vec{V}. \tag{7}$$

For two-dimensional nanofluids, Equation (7) takes the form:

$$\frac{\partial u}{\partial x} + \frac{\partial v}{\partial y} = 0 \tag{8}$$

2.2. Momentum Equation

The momentum equations for the modeled geometry and flow assumptions, are given by [74]:

$$\rho \frac{D}{Dt}(\vec{V}) = \nabla.\vec{V} + \vec{J} \times \vec{B} + \vec{g}. \tag{9}$$

Here ρ denotes the fluid density, $\vec{V}$ velocity, which can be expressed in components form as: $\vec{V} = (u, v, 0)$, $\vec{T}$ Cauchy tensor of stress and $\vec{g}$ force acts from a distance respectively. $\vec{J} \times \vec{B}$ represents the famous Lorentz force, in which $\vec{J}$ is the current density, B is the magnetic field with a magnetic field strength B_0. Furthermore, $\vec{J}$ can be expressed as: $\vec{J} = \sigma(\vec{E} + \vec{V} \times \vec{B})$, also known as Ohm's law, in which σ and $\vec{E}$ describe the electrical conductivity and electric field respectively, and assume that $\vec{E} = 0$. $\frac{D}{Dt}$ represents the substantial derivative. Using all the above assumptions in Equation (9), we get

$$\frac{\partial u}{\partial t} + u\frac{\partial u}{\partial x} + v\frac{\partial v}{\partial y} - v\frac{\partial^2 u}{\partial y^2} = \frac{\alpha_1}{\rho}\left(\frac{\partial}{\partial y}\left(u\frac{\partial^2 u}{\partial y^2}\right) - \frac{\partial u}{\partial y}\frac{\partial^2 u}{\partial x \partial y} + v\frac{\partial^2 u}{\partial y^2}\right) - \sigma B_0^2 u + g_r\beta_T(T - T_\infty) + g_r\beta_T(C - C_\infty), \tag{10}$$

where B_0 and α_1 describe the strength the magnetic field and thermal conductivity of the fluid, respectively, β_T is the thermal expansion coefficient, C_∞ and T_∞ denotes the concentration and temperature at a distance from the surface, and g is the gravitational acceleration.

2.3. Equation of Thermal Energy

Thermal energy equation for the unsteady flow field is presented in the form [74]:

$$\frac{\partial T}{\partial t} + \vec{V}.\nabla T = \nabla.\left[\frac{K(T)}{(\rho c_p)_f}\nabla T\right] + \tau\left[D_B\nabla C.\nabla T + \left(\frac{D_T}{T_0}\nabla T.\nabla T\right)\right], \tag{11}$$

where T represents the temperature, $\tau = \frac{\rho_p}{\rho_f}$ illustrates the ratio of the base liquid to thermal capacities of nanoparticles, the heat capacitance of the liquid is represented by c_p, D_B represents the constant of Brownian diffusion, while D_T demonstrates the constant of thermophoretic diffusion, T_0 denotes the liquid temperature, which is detached from the sheet.

After implementing the assumptions (two-dimensional, unsteady, viscous, in-compressible, electrically conducting, etc.), Equation (11) is reduced to:

$$\frac{\partial T}{\partial t} + u\frac{\partial T}{\partial x} + v\frac{\partial T}{\partial y} = \frac{1}{\rho c_p}\frac{\partial}{\partial y}\left[\frac{\partial u}{\partial y}\right] + \tau\left[D_B\left(\frac{\partial C}{\partial y}\frac{\partial T}{\partial y}\right) + \frac{D_T}{T_\infty}\left(\frac{\partial T}{\partial y}\right)^2\right]. \tag{12}$$

2.4. Equation of Mass Transfer

The nanofluid concentration can be mathematically described as [14]:

$$\frac{\partial C}{\partial t} + \vec{V}.\nabla C = D_B\nabla^2 C + \frac{D_T}{T_0}\nabla^2 T. \tag{13}$$

After applying the assumptions, Equation (13) takes the form:

$$\frac{\partial C}{\partial t} + u\frac{\partial C}{\partial x} + v\frac{\partial C}{\partial y} = D_B\frac{\partial^2 C}{\partial y^2} + \frac{D_T}{T_\infty}\left(\frac{\partial^2 T}{\partial y^2}\right). \tag{14}$$

The pertinent restrictions can be written as:

$$u = U_w, \quad v = 0, \quad T = T_w, \quad C = C_w \qquad \text{at} \quad y = 0, \tag{15}$$

$$\frac{\partial u}{\partial x} = \frac{\partial T}{\partial x} = \frac{\partial C}{\partial x} = 0, \quad v = \frac{dh(t)}{dt}, \quad C > 0, \qquad \text{at} \quad y = h. \tag{16}$$

Introducing the following transformations [14]:

$$\psi = x\sqrt{\frac{v\gamma}{1 - \zeta t}}f(\eta), \; u = \frac{\partial \psi}{\partial y} = \gamma x \frac{f'(\eta)}{1 - \zeta t}, \; v = \frac{\partial \psi}{\partial x} = -\sqrt{\frac{\gamma v}{(1 - \zeta t)}}f(\eta),$$

$$\eta = \sqrt{\frac{\gamma}{v(1 - \zeta t)}}y, \; h(t) = \left[\frac{v}{\gamma(1 - \zeta t)^{-1}}\right]^{\frac{1}{2}}, \; \theta(\eta) = \frac{T - T_0}{T_w - T_0}, \; \phi(\eta) = \frac{C - C_0}{C_w - C_0}. \tag{17}$$

Here the stream function is represented by ψ, the thickness of the fluid film is denoted by $h(t)$, and the kinematic viscosity is represented by $v = \frac{\mu}{\rho}$. The dimensionless film thickness is defined as:

$$\beta = \sqrt{\frac{\zeta}{v(1 - \zeta t)}}h(t) \tag{18}$$

In other words, Equation (18) becomes:

$$\frac{dh}{dt} = -\frac{\zeta \beta}{2}\sqrt{\frac{v}{\zeta(1 - \zeta t)}}. \tag{19}$$

With the help of the newly introduced variables, Equations (10)–(14) are reduced to the following equations, while the continuity equation is satisfied identically.

$$f''' + \gamma_1 \left(f''' f' - (f'')^2 - f f^{iv}\right) + f f'' - (f')^2 - St\left(f' + \frac{\eta}{2}f''\right) - Gr\theta + Gm\phi + Mf' = 0, \tag{20}$$

$$(1 + Rd)\theta'' + f\theta' - 2f'\theta - \frac{St}{2}(3\theta + \eta\theta') + Nt(\theta')^2 + Nb\theta'\phi' = 0, \tag{21}$$

$$\phi'' + Sc\left[f\phi' - 2f'\phi - \tfrac{St}{2}(3\phi + \eta\phi')\right] + \frac{Nt}{Nb}\theta'' = 0. \tag{22}$$

The boundary conditions of the problem are:

$$f(0) = 0, \; f'(0) = 1, \; \theta(0) = 1, \; \phi(0) = 1, \; f(\beta) = \frac{S\beta}{2}, \; f''(\beta) = 0, \; \theta'(\beta) = 0, \; \phi'(\beta) = 0. \tag{23}$$

After generalization, the physical parameters are defined as: $St = \frac{\chi}{\epsilon}$ is the measure of the time-dependent non-dimensional parameter, $\gamma_1 = \frac{\alpha_1 \beta^2}{\rho \delta^2}$ is the second-grade fluid stretching parameter, $M = \frac{\sigma_f B_0^2}{b\rho_f}$ represents the magnetic parameter, $\lambda = m(T - T_0)$ denotes the variable viscosity, $Pr = \frac{\rho v c_p}{k}$ represents the Prandtl number, $Nt = \frac{\tau D_w(T_w - T_\infty)}{vT_\infty}$ demonstrates thermophoresis parameter, $Nb = \frac{\tau D_B(C_w - C_\infty)}{v}$ represents the limitations of Brownian motion, $Gm = \frac{g\beta L^3(C_w - C_\infty)}{v}$ is the mixing parameter, $Gr = \frac{g\beta L^3(T_w - T_\infty)}{v}$ denotes the Grashof number, $Rd = \frac{16\sigma I_\infty^3}{3kk^*}$ is the radiation parameter, and $Sc = \frac{v}{D_B}$ represents Schmidt number.

3. Parameters of Interest

3.1. Skin Friction

The coefficient of skin friction can be defined in the closed form as:

$$C_f = \frac{(\vec{S_{xy}})_{y=0}}{\frac{\rho U_w^2}{2}},\tag{24}$$

where

$$\vec{S}_{xy} = \left[\mu \frac{\partial u}{\partial y} + \rho a_1 \left(2 \frac{\partial u}{\partial x} \frac{\partial u}{\partial y} + \frac{\partial^2 u}{\partial x \partial y} \right) \right]\tag{25}$$

A dimensionless description of C_f is demonstrated as:

$$C_f = (Re_e)^{-\frac{1}{2}} (f''(0) + 3\gamma_1 f''(0) f'(0)),\tag{26}$$

where R_e denotes the local Reynolds number, and has the mathematical description given as: $R_e = \frac{U_w x}{\nu}$.

3.2. Nusselt Number

Nusselt number has the closed mathematical form given by $Nu = \frac{hQ_w}{k(T_0 - T_h)}$, where $Q_w = -\hat{k} \left(\frac{\partial T}{\partial y} \right)_y = 0$, and is known as the flux of heat. A dimensionless description of Nu is demonstrated as:

$$Nu = -\Theta(0)\tag{27}$$

3.3. Sherwood Number

Sherwood number can be demonstrated in mathematical form as: $Sh = \frac{hJ_w}{D_B(C_0 - C_h)}$, where $J_w = -D_B \left(\frac{\partial C}{\partial y} \right)_{y=0}$ is the flux of mass. The non-dimensional descriptions of Sh is demonstrated as:

$$Sh = -\Phi(0)\tag{28}$$

4. Entropy Analysis and Its Mathematical Description

Entropy generation of volumetric type of viscous fluids is demonstrated as [87,88]:

$$S''' = \frac{K(T)}{T_0^2} \left[\left(\frac{\partial T}{\partial y} \right)^2 + \frac{16\sigma T_\infty^3}{3k} \left(\frac{\partial T}{\partial y} \right)^2 \right] + \frac{\mu(T)}{T_0} \left(\frac{\partial u}{\partial y} \right)^2 + \frac{Rd}{C_0} \left(\frac{\partial C}{\partial y} \right)^2 + \frac{Rd}{T_0} \left(\frac{\partial T}{\partial y} \frac{\partial C}{\partial y} + \frac{\partial C}{\partial x} \frac{\partial T}{\partial x} \right) + \frac{\sigma B_0^2 u^2}{T_0}\tag{29}$$

Equation (29) illustrates that the entropy generation has two main features, i.e., the irreversibility of the transmission of heat, and the fluid friction irreversibility. Magnetic and porosity effects are illustrated in the last two terms. The aspects of entropy generation are illustrated by mathematical description, given by

$$S_0''' = \frac{K_0 (\Delta T)^2}{L^2 T_0^2},\tag{30}$$

while $N_G = \frac{S'''}{S_0'''}$, demonstrates the ratio of the actual entropy generation to the generation rate of characteristic entropy. This number N_G for a non-dimensional system takes the form:

$$N_G = R_e (1 + \epsilon \theta + Rd)(\theta')^2 + \frac{R_e Br}{\Omega(1 + \Lambda)} (f'')^2 + \frac{R_e Br}{\Omega} M(f')^2 + R_e \lambda \left(\frac{\chi}{\Omega} \right)^2 (\phi')^2 + R_e \lambda \left(\frac{\chi}{\Omega} \right) \theta' \phi',\tag{31}$$

where Br is used for the Brinkman number, λ represents the diffusion quantity, M demonstrates the parameter of the magnetic field, Ω and χ denotes the dimensionless temperature and concentration change, respectively.

The illustrated mathematical description of each parameter is given as:

$$R_e = \frac{bL^2}{\nu}, \Omega = \frac{\Delta T}{T_0}, \chi = \frac{\Delta C}{C_0}, \lambda = \frac{RdC_0}{k}. \tag{32}$$

The source of the generation of the entropy is an important class for engineers, the Bejan number is responsible for such an important measurement, which is defined as:

$$B_e = \frac{\frac{K(T)}{T_0^2}\left[\left(\frac{\partial T}{\partial y}\right)^2 + \frac{16\sigma T_\infty^3}{3k}\left(\frac{\partial T}{\partial y}\right)^2\right]}{\frac{\mu(T)}{T}\left(\frac{\partial u}{\partial y}\right)^2 + \frac{\sigma B_0^2 u^2}{T}} \tag{33}$$

The alternative representation of the Bejan number B_e, after using the similarity transformations, is:

$$B_e = \frac{(1 + \epsilon\theta + Rd)(\theta')^2}{\frac{Br}{\Omega(1+\Lambda)}(f'')^2 + \frac{Br}{\Omega}M(f')^2} \tag{34}$$

5. Solution by HAM

Numerical methods applied presently uses the concept of linearization and discretization to tackle the nonlinear systems. HAM is the method used for the same purpose, analytically, as the numerical techniques. Its derivation is totally dependent on the topological concept known as homotopy. Liao [97,98] was the first to use this concept successfully. For this purpose, Liao used the idea of homotopy by considering two continuous functions Ψ_1 and Ψ_2 defined on the topological spaces $\bar{X}$ and $\bar{Y}$. The homotopic idea explained in topological spaces is implemented over a closed unit interval by:

$$\Psi : \bar{X} \times [0,1] \to \bar{Y}$$

where the relation holds for all $\bar{x} \in \bar{X}$, together with $\Psi[\hat{x},0] = \zeta_1(\hat{x})$ and $\Psi[\hat{x},1] = \zeta_2(\hat{x})$. The mapping defined by Ψ is known as a homotopic function in the literature. Equations (20)–(23) have been tackled by HAM. Solutions are restricted by a new parameter, $\hbar$, which modifies and links the solutions in an appropriate manner.

We guess

$$f_0(\eta) = \frac{1}{4\beta^2}(4\beta^2\eta + 3\beta(S-2)\eta^2 + (2-S)\eta^3), \quad \theta_0(\eta) = 1, \quad \phi_0(\eta) = 1. \tag{35}$$

The linear operators represented by $L_{\tilde{f}}$, $L_{\tilde{\theta}}$ and $L_{\tilde{\phi}}$ are defined as:

$$L_{\tilde{f}}(\tilde{f}) = \tilde{f}''', \quad L_{\tilde{\theta}}(\tilde{\theta}) = \tilde{\theta}'' \quad \text{and} \quad L_{\tilde{\phi}}(\tilde{\phi}) = \tilde{\phi}'', \tag{36}$$

with the following properties

$$L_{\tilde{f}}(a_1 + a_2\eta + a_3\eta^2) = 0, \quad L_{\tilde{\theta}}(a_4 + a_5\eta) = 0, \quad \text{and} \quad L_{\tilde{\phi}}(a_6 + a_7\eta) = 0. \tag{37}$$

6. HAM Solution Convergence

The goal of the HAM solution (series solution) is its convergence, which is always linked to some constraints. The subordinate restrictions $\hbar_{\tilde{f}}$, $\hbar_{\tilde{\theta}}$, and $\hbar_{\tilde{\phi}}$ are implemented in the HAM procedure. The embedding parameters choice guarantee the solution convergence [99]. In our case, the proposed method shows the performance in the form of results, which are valid and efficient. The probability sectors of $\hbar$ are constructed $\hbar$-curves of $\tilde{f}''(0)$, $\tilde{\theta}'(0)$ and $\tilde{\phi}'(0)$ for the HAM solution (approximated) of order 25. The effective domains of $\hbar$ are $-1.5 < \hbar_{\tilde{f}} < 0.0$, $-1.5 < \hbar_{\tilde{\theta}} < 0.0$ and $-2.5 < \hbar_{\tilde{\phi}} < 0.0$. Figure 1 illustrates the HAM technique convergence of $\hbar$-curves for the state variables (velocity, temperature, and concentration). The numerical values of HAM solution convergence with the variations of different parameters are presented in Tables 1 and 2. The table illustrates that HAM is a rapidly convergent method.

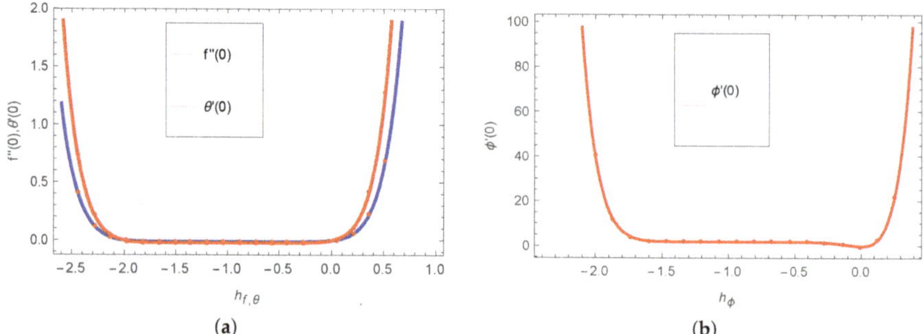

(a) (b)

Figure 1. (**a**,**b**) $\hbar$-curve representation of $f(\eta)$, $\theta(\eta)$ and $\phi(\eta)$.

Table 1. 25*th*-order approximation table for HAM convergence, when $\gamma_1 = Sc = 0.5$, $Nt = Gr = \beta = St = 0.1$, $h = -0.5$, $Pr = Nb = Gm = 0.1$.

Approximation Order	$f''(0)$	$\Theta'(0)$	$\phi'(0)$
1	-0.0600000	-1.107500	0.0537500
3	-0.104656	-0.106980	0.0856833
5	-0.115692	-0.106849	0.0935671
7	-0.118420	-0.106817	0.0955156
9	-0.119090	-0.106809	0.0095997
11	-0.119261	-0.106807	0.0961163
13	-0.119303	-0.106807	0.0961457
15	-0.119313	-0.106806	0.0961530
17	-0.119315	-0.106806	0.0961548
19	-0.119316	-0.106806	0.0961552
23	-0.119316	-0.106806	0.0961554
25	-0.119316	-0.106806	0.0961554

7. Results and Discussion

The objective of our investigation focuses on the interpretation of the thin film flow of nanoliquid flow parameters. Figures 2–23 reveal the comprehension of the parameters involved in our model equations.

7.1. Velocity Profile

The liquid film thickness β along the direction of the fluid flow is illustrated in Figure 2. The flow velocity curve declines with the larger numbers of the film thickness β, because the dimensionless thin film thickness is directly related to the fluid thickness $h(t)$ and is the function of viscosity. As a result, an increase in β further increases the viscosity of the film, which further causes in the decline of the velocity curve. This happens due to the indirect relationship between β and the flow velocity profile, i.e., the increasing values of β decrease the viscosity of the fluid, which as a result decreases the velocity profile.

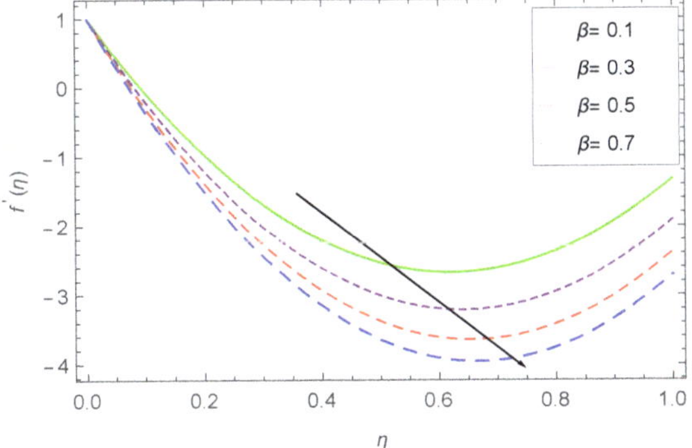

Figure 2. Impact of β on $f'(\eta)$, when $\gamma_1 = 0.2$, $St = 0.6$, $Gm = 0.8$, $Gr = 0.7$, $M = 0.8$.

Figure 3 reflects the impact of the unsteadiness parameter over the profile of the velocity for dissimilar values of the embedded parameters. A direct variation can be observed in the profile of the velocity with the variations in the unsteadiness parameter St in the figure as demonstrated. These variations are due to the effect of the stretching parameter. The unsteadiness parameter is a function of the liquid film thickness, which further varies directly with the stretching parameter and as a result increases the velocity profile. An increase in St rises the motion of the fluid. Investigation demonstrates that the solution exists for $St \in [0, 2]$ and strongly depends on the parameter St.

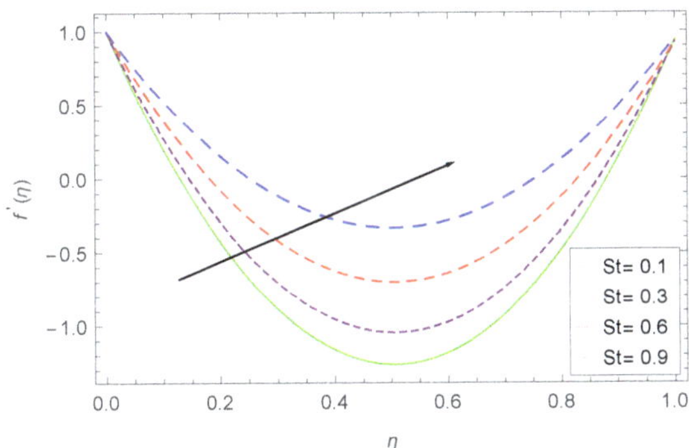

Figure 3. Impact of St on $f'(\eta)$, when $\gamma_1 = 0.2$, $\beta = 0.9$, $Gm = 0.8$, $Gr = 0.7$, $M = 0.7$.

Figure 4 illustrates the mixed convective effect on the flow. In Figure 4 Gm shows the characteristic of buoyancy forces, which play a favorable behavior for the state variable velocity. The mixing parameter is mainly affected by the length, concentration difference, and the kinematic viscosity of the nanofluid. There is an inverse relation between the velocity and the viscosity of the nanofluid. Therefore, when the mixing parameter increases, the liquid film concentration increases directly, and the viscosity decreases, which further causes increase of the profile of the velocity. The velocity profile shows a direct relation to Gm.

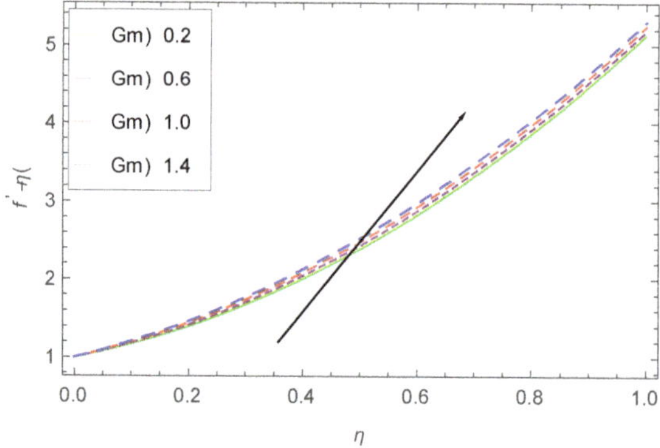

Figure 4. Impact of *Gm* on $f'(\eta)$, when $\gamma_1 = 0.2$, $\beta = 0.9$, $St = 0.3$, $Gr = 0.7$, $M = 0.9$.

Figure 5 reveals the effect of the order two-fluid velocity distribution parameter γ_1. Physically, this parameter is inversely related to the density, keeping the thickness parameter constant. Therefore, an increase in the values of the parameter γ_1, would decrease the density of the fluid, which further causes in the increase of the velocity profile. In other words, it would make the fluid less dense and as a result the fluid velocity jumps up. Thus, the larger the values of non-Newtonian parameter, the greater the motion of thin film.

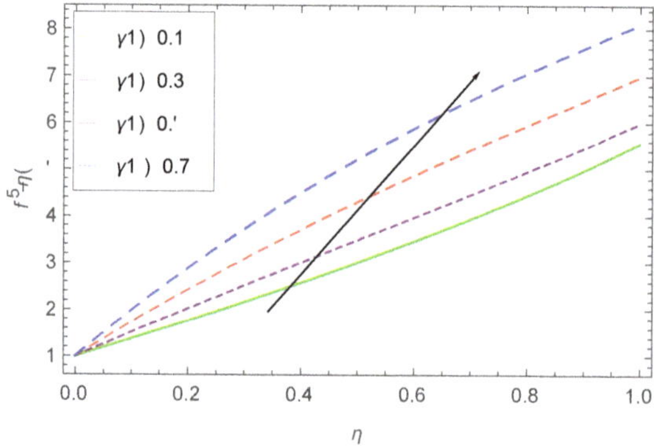

Figure 5. Impact of γ_1 on $f'(\eta)$, when $Gm = 0.8$, $\beta = 0.9$, $St = 0.3$, $Gr = 0.7$, $M = 0.7$.

Figure 6 illustrates the relation between Grashof parameter *Gr* and the profile of the velocity. Here *Gr* shows the characteristics of buoyancy forces, which offers a favorable behavior for the velocity profile. Physically, Grashof number *Gr* is the ratio of the buoyancy force to the viscous force. Increasing values of the buoyancy forces cause the decrease of the viscous forces, which as a result producing faster motion. In summary, the increasing values of *Gr* causes a rapid increase in the velocity profile.

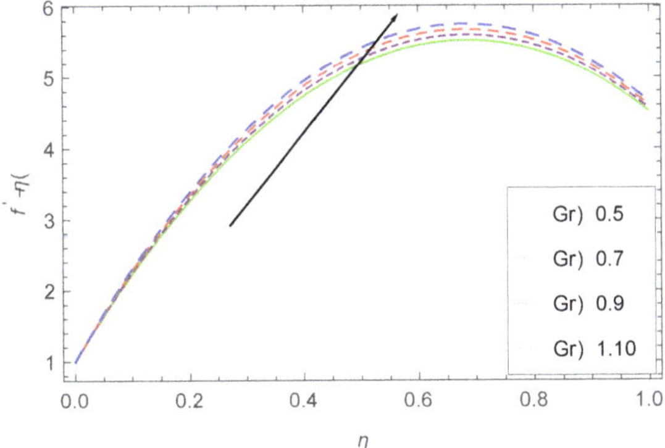

Figure 6. Impact of Gr on $f'(\eta)$, for $\gamma_1 = 0.4$, $\beta = 0.6$, $St = 0.5$, $Gm = 0.8$, $M = 0.7$.

Figure 7 illustrates the variation of the magnetic parameter M over the velocity profile. Since the magnetic parameter is applied horizontally to the surface on which the nanofluid flows, so an increase in the magnetic parameter would increase the strength of the magnetic field that create bending on the surface of the plate. This bending causes the decline of the velocity profile, but does nothing to the magnitude. In short, with the increasing values of the magnetic parameter M, a decrease is observed in the velocity profile.

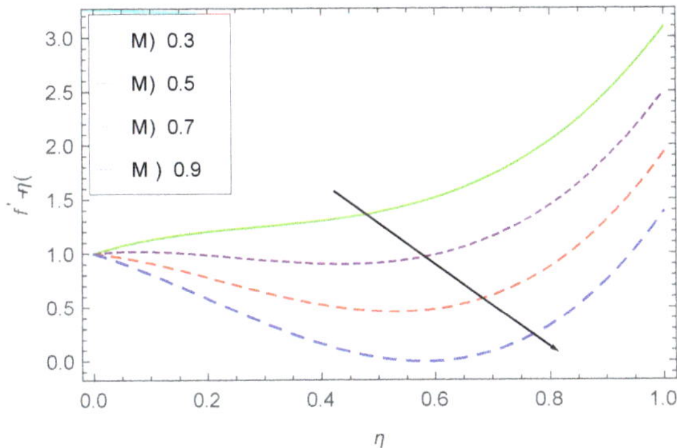

Figure 7. Impact of M on $f'(\eta)$, for $\gamma_1 = 0.2$, $\beta = 0.6$, $St = 0.6$, $Gr = 0.5$, $Gm = 0.8$.

7.2. Temperature Profile

The effect of thermal radiations Rd on $\theta(\eta)$ is discussed in Figure 8. The figure reveals an inverse relationship between the thermal radiation parameter Rd and the temperature profile. For higher values of Rd a rapid decrease can be observed in the profile of the temperature and vice versa.

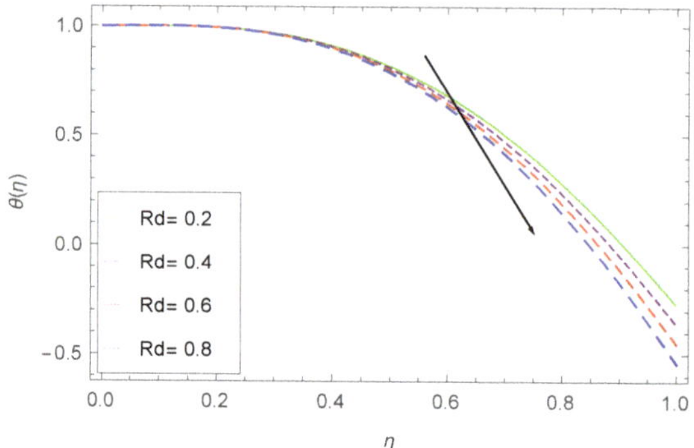

Figure 8. Impact of Rd on $\theta(\eta)$, for $\beta = 0.1$, $St = 0.1$, $Nt = 0.6$, $Nb = 0.5$.

Figure 9 reveals the impact on the temperature profile of the thermophoresis parameter Nt. The limitations on thermophoresis helps in the increase of a surface temperature. The irregularity in motion (Brownian motion) causes a temperature increase due to the kinetic energy produced by nano-suspended particles; consequently, a thermophoretic force is produced. The fluid starts in the opposite direction of the stretching sheet, due to the intensity produced by this force. As a result, larger values of Nt cause an increase in temperature, due to which the surface temperature also increases.

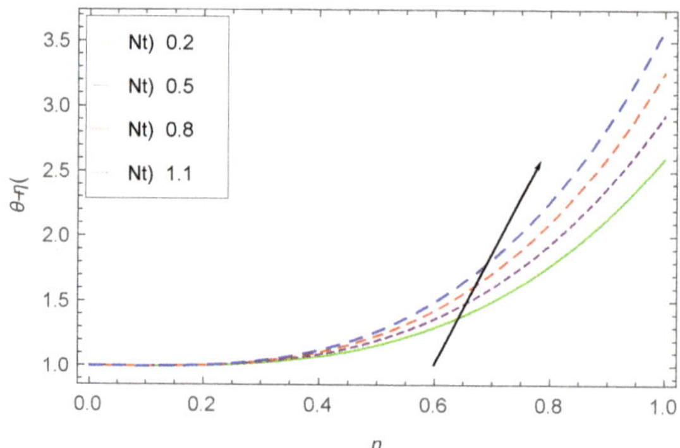

Figure 9. Impact of Nt on $\theta(\eta)$, for $\beta = 0.1$, $St = 0.1$, $Rd = 0.2$, $Nb = 0.3$.

Figure 10 demonstrates the effect on the profile of a heat $\theta(\eta)$ of the unsteadiness parameter St. It is observed that $\theta(\eta)$ directly varies with unsteadiness parameter St. The higher numbers of the unsteadiness parameter St increases the temperature, which further causes increase in the kinetic energy of the fluid, the result of which appears in the form of increase in the liquid film.

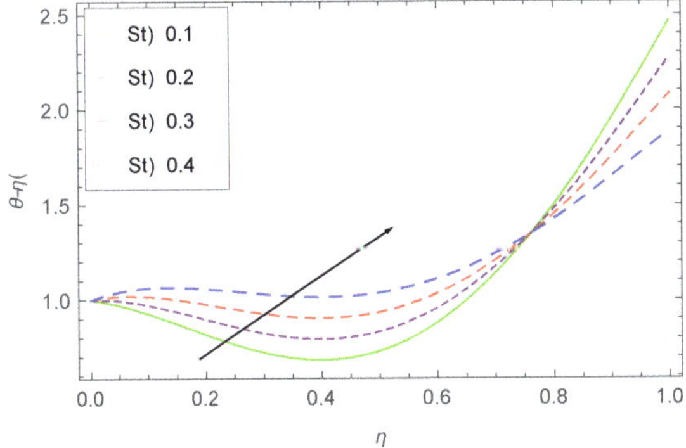

Figure 10. Impact of St on $\theta(\eta)$, for $\beta = 0.1$, $Nt = 0.8$, $Rd = 0.2$, $Nb = 0.5$.

The impact of thin film thickness β on temperature for different values of the embedded parameter is presented in Figure 11. Since the thickness parameter is the function of the kinematic viscosity and fluid thickness, so an increase in β would increase the viscosity, which further causes the decline of the temperature profile. Thus, for larger values of β, the profile of the temperature falls. The same effect can be seen in the profile of the velocity for β.

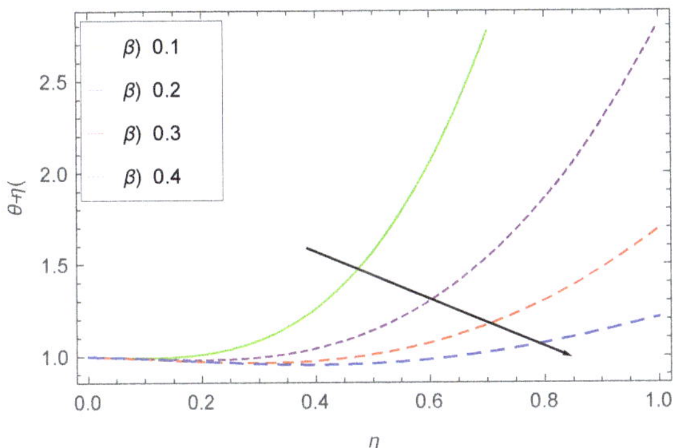

Figure 11. Impact of β on $\theta(\eta)$, for $St = 0.1$, $Nt = 0.5$, $Rd = 0.6$, $Nb = 0.5$.

Figure 12 illustrates the temperature distribution under Brownian motion parameter Nb. In general, due to the irregular motion of particles, a collision is produced between the particles. The figure shows that an increase in heat of the fluid can be observed with the ascending order of the Brownian motion parameter, Nb; consequently, free surface nanoparticle volume friction decreases.

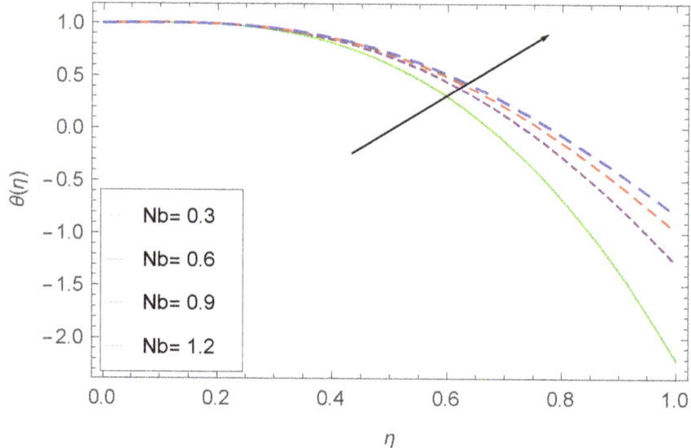

Figure 12. Impact of Nb on $\theta(\eta)$, for $St = 0.2$, $Nt = 0.6$, $Rd = 0.2$, $\beta = 0.1$.

7.3. Concentration Profile

Figure 13 illustrates the effects on concentration profile $\phi(\eta)$ of the Brownian motion parameter Nb. The irregularity and turbulence in the motion of the fluid particles is normally known as Brownian motion. At molecular level, Brownian motion of micropolar nanofluid leads to the thermal conductivity of nanofluids. The figure illustrates the increase in Nb in the form of a decline in the profile of the concentration. The boundary-layer thicknesses diminish due to the larger values of Brownian motion, which results in reduction of the concentration profile.

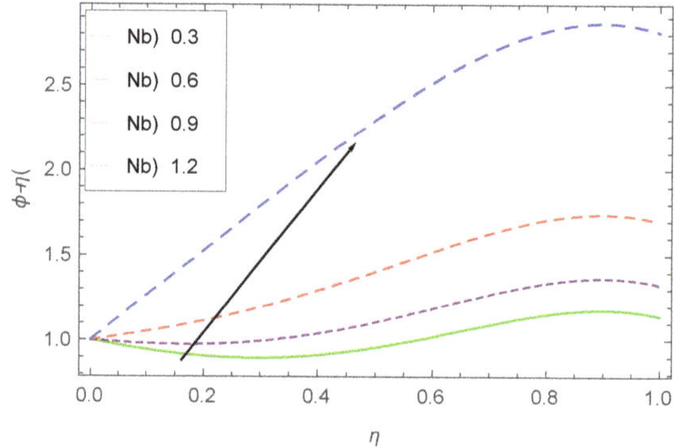

Figure 13. Impact of Nb on $\phi(\eta)$, for $St = 0.1$, $Nt = 0.1$, $Sc = 0.5$, and $\beta = 0.5$.

Figure 14 shows the concentration profile $\phi(\eta)$ behavior, under the effect of the unsteadiness parameter St. A direct relation can be observed between the unsteadiness parameter St and the concentration profile $\phi(\eta)$. Higher values of the unsteadiness parameter St increases the profile of temperature that blows the kinetic energy of the fluid, which further causes an increase in the concentration of the liquid film.

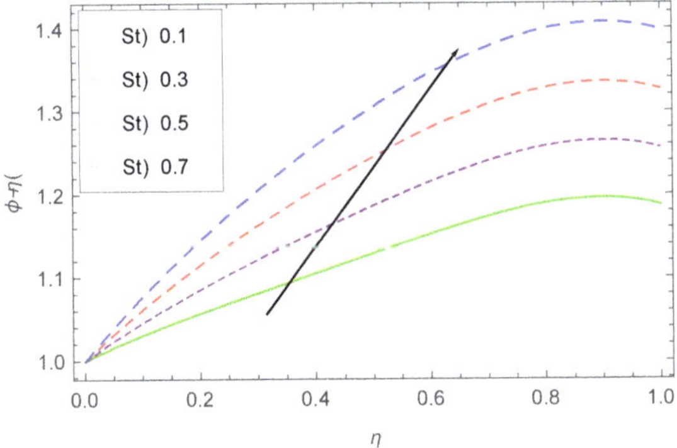

Figure 14. Impact of St on $\phi(\eta)$, for $Nt = 0.1$, $Nb = 1.2$, $Sc = 0.5$, and $\beta = 0.9$.

Figure 15 illustrates the effect on concentration field of the thermophoresis parameter Nt. The figure demonstrates that the concentration profile rises due to an increase in Nt. This is because of higher values of Nt increase the nanofluid molecule kinetic energy, as a result of which the concentration increases.

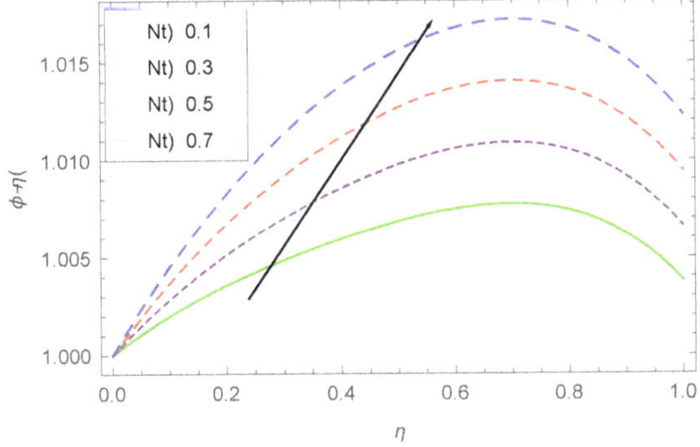

Figure 15. Impact of Nt on $\phi(\eta)$, when $St = 0.1$, $Sc = 0.5$, $Nb = 1.2$, $\beta = 0.9$.

Figure 16 illustrates the thin film thickness β effect on $\phi(\eta)$ for different values of the embedded parameters. The thickness parameter is inversely related to the kinematic viscosity; we know that kinematic viscosity is inversely proportional to the density of the fluid, so increasing the thickness parameter β causes the decline of the concentration profile. Thus, it is obvious that the concentration profile falls with higher numbers of β. The same effect was observed for β in the velocity distribution as well as in temperature distribution.

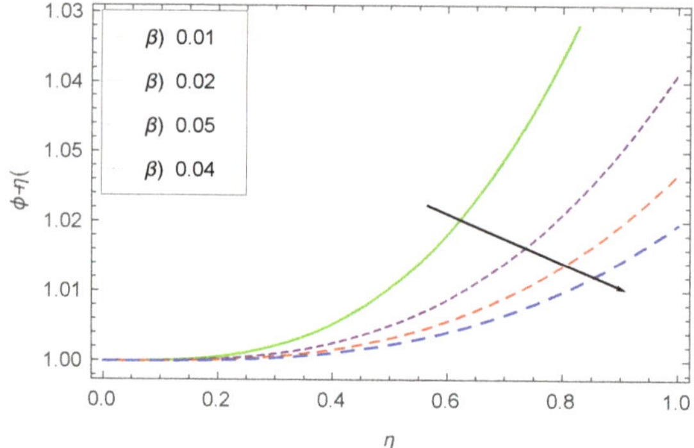

Figure 16. Impact of β on $\phi(\eta)$, when $Nt = 0.1$, $Nb = 0.3$, $Sc = 0.3$, $St = 0.1$.

Figure 17 demonstrates the opposite information as discussed in the temperature distribution under different parameters. The diagram shows that with increasing values of Schmidt number Sc, the concentration profile decreases, consequently reducing the thickness of the boundary-layer.

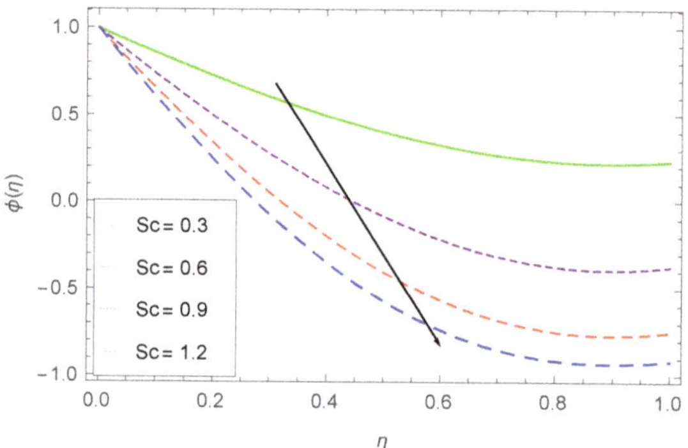

Figure 17. Impact of Sc on $\phi(\eta)$, for $Nt = 0.1$, $Nb = 0.3$, $\beta = 0.9$, $St = 0.1$.

7.4. Entropy Profile

The diagram reflects the variations of Brinkman number Br verses Bejan number B_e. It is observed that with increasing values of Br the Bejan number B_e declines. There is no variation in the Bejan number, and it remains constant up to 0.8 for the values of η in decreasing order for different values of the Brinkman number. Large variations are observed for smaller values of η as we increase the values of Brinkman number.

Figure 18 illustrates the variations in Bejan number under dissimilar values of the magnetic parameter M. It is clear from the figure that for larger values of M, the Bejan number boots up. Similar variations are observed for Brinkman number in Figure 19. The variations here are constant for smaller values of the magnetic parameter in the range of $0.85 < \eta \leq 1.0$. Large variations are investigated for greater values of the magnetic parameter in the range of $0.0 < \eta \leq 0.2$.

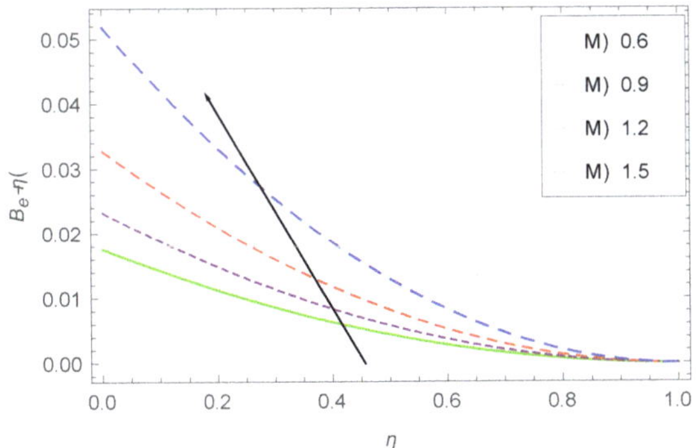

Figure 18. Deviations in Bejan number B_e for dissimilar numbers of M, when $Br = 0.1$, $R_e = 0.7$, $Rd = 0.6$, and $Pr = 0.6$.

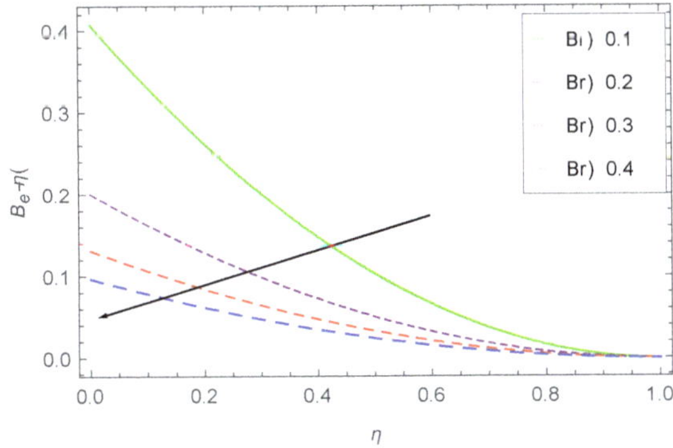

Figure 19. Variations in B_e for dissimilar values of Br, when $M = 0.1$, $R_e = 0.7$, $Rd = 0.6$, and $Pr = 0.6$.

Figure 20 illustrates the effect of Reynolds number on $N_G(\eta)$. For greater values of R_e the entropy regime increases. Consequently, Reynolds number and entropy function variates directly. From the graph it is clear that for large values of η the variations in entropy remains constant, while this variation jumps up for large values of the Reynolds number in the range of $0.0 \leq \eta \leq 0.4$.

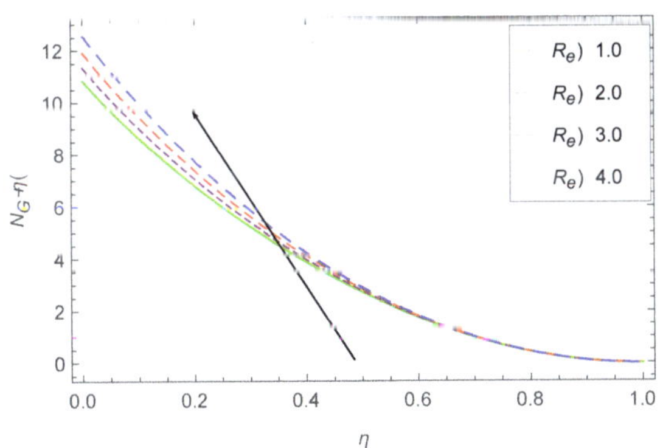

Figure 20. Variations in the function of entropy $N_G(\eta)$ for dissimilar values of R_e, when $Br = 0.1$, $Rd = 0.6$, $M = 1.3$.

Figure 21 demonstrates the variations in entropy number with the variations in Prandtl number Pr. The figure demonstrates that for larger values of the Prandtl number Pr, the temperature profile rises, and consequently, the entropy function boots up. The variations in the entropy function jumps up throughout inside the interval $0.0 \leq \eta \leq 1.0$. The increasing values of the Prandtl number increases the entropy function exponentially.

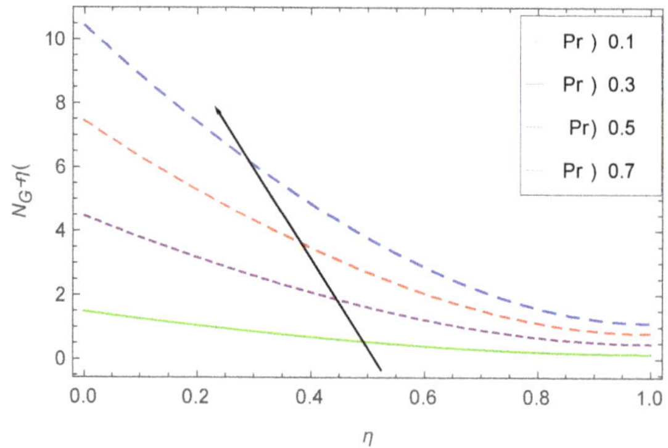

Figure 21. Variations in the function of entropy $N_G(\eta)$ for unlike values of Pr, when $Br = 0.3$, $R_e = 0.7$, $Rd = 0.7$ and $M = 0.9$.

Figure 22 demonstrates the variations of magnetic parameter on the entropy function $N_G(\eta)$. It is clear from the figure that for larger numbers of M, the entropy function reduces, because the Lorentz force in the magnetic field produces the resistance strength. These variations in the entropy function exist in the range of $0.0 \leq \eta \leq 0.65$ for larger values of the magnetic parameter, and remains constant elsewhere.

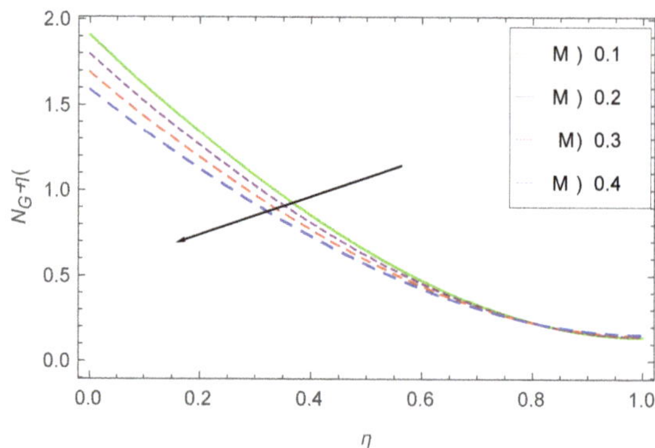

Figure 22. Variations in the function of entropy $N_G(\eta)$ for unlike values of M, when $Br = 0.3$, $R_e = 0.7$, $Rd = 0.7$ and $Pr = 1.2$.

Figure 23 demonstrates the effect of Brickman number Br on the generation function of entropy generation $N_G(\eta)$. Figure shows that for larger numbers of the Brickman number $N_G(\eta)$ increases. Viscous dissipation produces heat, which as a result raises the generation of entropy due to the lower conduction rate. The figure demonstrates that the variation in the entropy function remains constant between $0.64 < \eta \leq 1.0$, while an increasing tendency is observed for large values of the Brinkman number elsewhere.

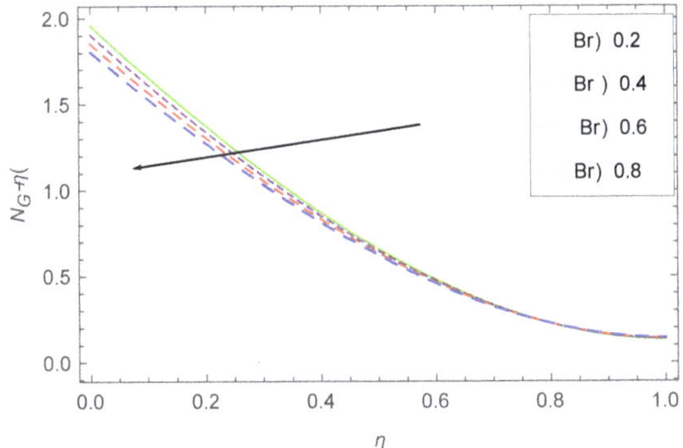

Figure 23. Variations in the function of entropy $N_G(\eta)$ for unlike values of Br, when $M = 0.7$, $R_e = 0.6$, $Rd = 0.8$ and $Pr = 1.5$.

7.5. Table Discussion

The influence of different parameters is demonstrated in Tables 2–4. Table 2 explains the effect on skin friction of M, Gr, β, and St. The skin-friction coefficient shows a rapid increase, with larger numbers of the unsteady parameter St, while for large values of the magnetic parameter M the Nusselt number decreases. On the other hand, larger values of the Grashof number Gr and thickness parameter β cause the decline of the skin-friction coefficient. Physically, Grashof number arises in the natural convection due to density difference. Viscous forces are also the functions of dependence of the Grashof number. Increasing the viscosity coefficient causes a decrease in the Grashof number. Therefore, as a result, the increasing values reduce the skin friction.

Table 3 reveals the effect on Nu of the parameters M, St, β, and Pr. It is noted that the larger the values of M (magnetic parameter), and the unsteadiness parameter St, the smaller the Nusselt number Nu. On the other hand, the Nusselt number Nu increases with larger values of unsteadiness parameter St and thickness parameter β. The increasing values of the Prandtl number decreases Nu.

The effects of Brownian motion Nb, thermophoretic parameter Nt, Schmidt number Sc, Prandlt Pr, and unsteadiness parameter St on Sherwood number are demonstrated in Table 4. It is clear that the increasing values of the thermophoretic parameter increases the Sherwood number. An inverse relation is observed between the Schmidt number and Sherwood number. The same phenomenon appears in observations for the unsteady parameter St and Prandtl number Pr for the Sherwood number in the form of decline. The larger values of the Prandtl number show an exponential decline in the Sherwood number.

Table 2. Variation in skin-friction coefficent, under dissimilar values of M, Gr, β and St.

M	Gr	β	St	C_f
0.1	0.5	0.1	1.5	−0.1708
0.5				−0.2034
1.0				−0.2460
1.5	0.1			−0.1776
	0.5			−0.1708
	1.0			−0.1708
	1.5	0.1		0.3434
		0.5		0.7463
		1.0		−6.9860
		1.5	0.1	−1.8672
			0.5	−0.2217
			1.0	−0.1320

Table 3. For dissimilar values of *Pr*, *M*, *Rd* and *St* variations in *Nu* (Nusselt Number).

M	*β*	*St*	*Pr*	*Nu*
0.1	0.1	1.5	1.5	0.3523
0.5				0.3521
1.0				0.3519
1.5	0.1			0.3523
	0.5			1.5145
	1.0			2.2616
	1.5	0.1		1.6248
		0.5		0.2284
		1.0		0.2904
		1.5	1.5	0.3523
			3.0	0.3243
			5.0	0.3087
			7.0	0.3012

Table 4. For dissimilar values of *Nb*, *Nt*, *Sc*, *Pr*, and *St*, variations in Sherwood Number.

Nb	*Nt*	*Sc*	*St*	*Pr*	*Sh*
0.1	0.5	0.1	1.5	1.5	−1.3582
0.5					−0.2388
1.0					−0.0988
1.5	0.1				0.0223
	0.5				−1.3582
	1.0				−2.7537
	1.5	0.1			−4.1455
		0.5			−3.9667
		1.0			−3.7453
		1.5	0.1		−1.8105
			0.5		−2.3059
			1.0		−2.9193
			1.5	1.5	−3.5263
				3.0	−5.6130
				5.0	−6.7706
				7.0	−7.3323

8. Conclusions

These reconsideration efforts are worthy enough to categorize the enhancement in heat transfer and thermal conductivity of non-Newtonian fluid with nanoparticle conductive properties. The velocity profile shows an increase with increasing value of the unsteadiness parameter *St*, while the increasing values of the magnetic parameter causes the decline of the velocity profile of the nanofluid film. It is shown that the coefficient of skin friction rises with the larger rates of the magnetic parameter *M* and the unsteadiness parameter *St*; on the other hand, the coefficient of skin friction decreases with higher values of the stretching and thickness parameters. The temperature profile shows a direct variation with Brownian motion parameter. The thermal boundary-layer thickness decreases with increasing values of the of *Sc*. Nusselt number with increasing values of the radiation parameter increases. The surface temperature of the fluid increases with increasing values of Prandtl number, while an opposite tendency is observed with larger values of the unsteady parameter on the temperature profile. Similar results are investigated for the temperature profile with the variation of the thermophoresis parameter. The mass flux shows a decline with higher numbers of the Brownian motion parameter, while an opposite trend is experienced for the thermophoretic parameter. The implemented technique convergence is shown numerically for the validation of our technique.

Author Contributions: A.U., Z.S. and S.I. modeled the problem and wrote the manuscript. P.K. and M.A. thoroughly checked the mathematical modeling and English corrections. M.J. and A.U. solved the problem using

Mathematica software, and P.K., Z.S., M.J., and M.A. contributed to the results and discussions. All authors finalized the manuscript after its internal evaluation.

Funding: This research was funded by the Center of Excellence in Theoretical and Computational Science (TaCS-CoE), KMUTT.

Acknowledgments: This project was supported by the Theoretical and Computational Science (TaCS) Center under Computational and Applied Science for Smart Innovation Research Cluster (CLASSIC), Faculty of Science, KMUTT.

Conflicts of Interest: The authors declare no conflict of interest.

Abbreviations

The following abbreviations and parameters with their possible dimensions stated here are used in this article:

Sh	Sherwood number
β	Film thickness parameter
Nu	Nusselt number
St	Unsteady parameter
R_e	Reynolds number
Pr	Prandtl number
ζ	Stretching parameter
Sc	Schmidt number
U_w	Stretching velocity $\left(\frac{m}{sec}\right)$
Nt	Thermophoretic parameter
C_f	Skin-friction coefficient
Nb	Brownian motion parameter
T	Cauchy stress tensor
T	Fluid temperature (K)
I	Identity tensor
v	Kinematic viscosity $\frac{m^2}{sec}$
ρ	Density $\left(\frac{Kg}{m^3}\right)$
μ	Dynamic viscosity mPa
c_p	Specific heat $\left(\frac{J}{KgK}\right)$
$h(t)$	Thickness of liquid
Q_w	Heat Flux $\left(\frac{W}{m^2}\right)$
J_w	Mass flux $\left(\frac{Kg}{sec.m^2}\right)$
f	Dimensionless velocity
∞	Condition at infinity
0	Reference condition
$\tilde{u}$	Velocity component in $x-$direction $\left(\frac{m}{sec}\right)$
$\tilde{v}$	Velocity component in $y-$direction $\left(\frac{m}{sec}\right)$
x, y, z	Coordinates (m)
η	Similarity variable
t	Time (sec)
ω	Frequency parameter
A_1, and A_2	Revilin Erickson Tensor
E	Electric Field
α	Non-Newtonian Parameter
u_0	Magnetic Permeability
α_1 and α_2	Material constants
Gm	Mixing parameter
σ	Electric conductivity
J	Current density

References

1. Myers, T. Application of non-Newtonian models to thin film flow. *Phys. Rev. E* **2005**, *72*, 066302. [CrossRef] [PubMed]
2. Marinca, V.; Herişanu, N.; Nemeş, I. Optimal homotopy asymptotic method with application to thin film flow. *Open Phys.* **2008**, *6*, 648–653. [CrossRef]
3. Sandeep, N.; Malvandi, A. Enhanced heat transfer in liquid thin film flow of non-Newtonian nanofluids embedded with graphene nanoparticles. *Adv. Powder Technol.* **2016**, *27*, 2448–2456. [CrossRef]
4. Wang, C. Liquid film on an unsteady stretching surface. *Q. Appl. Math.* **1990**, *48*, 601–610.
5. Usha, R.; Sridharan, R. On the motion of a liquid film on an unsteady stretching surface. *ASME Fluids Eng.* **1993**, *150*, 43–48. [CrossRef]
6. Liu, I.C.; Andersson, H.I. Heat transfer in a liquid film on an unsteady stretching sheet. *Int. J. Therm. Sci.* **2008**, *47*, 766–772. [CrossRef]
7. Aziz, R.C.; Hashim, I.; Alomari, A. Thin film flow and heat transfer on an unsteady stretching sheet with internal heating. *Meccanica* **2011**, *46*, 349–357. [CrossRef]
8. Tawade, L.; Abel, M.; Metri, P.G.; Koti, A. Thin film flow and heat transfer over an unsteady stretching sheet with thermal radiation, internal heating in presence of external magnetic field. *Int. J. Adv. Appl. Math. Mech.* **2016**, *3*, e40.
9. Andersson, H.I.; Aarseth, J.B.; Dandapat, B.S. Heat transfer in a liquid film on an unsteady stretching surface. *Int. J. Heat Mass Transf.* **2000**, *43*, 69–74. [CrossRef]
10. Chen, C.H. Heat transfer in a power-law fluid film over a unsteady stretching sheet. *Heat Mass Transf.* **2003**, *39*, 791–796. [CrossRef]
11. Chen, C.H. Effect of viscous dissipation on heat transfer in a non-Newtonian liquid film over an unsteady stretching sheet. *J. Non-Newton. Fluid Mech.* **2006**, *135*, 128–135. [CrossRef]
12. Ishaq, M.; Ali, G.; Shah, S.I.A.; Shah, Z.; Muhammad, S.; Hussain, S.A. Nanofluid Film Flow of Eyring Powell Fluid with Magneto Hydrodynamic Effect on Unsteady Porous Stretching Sheet. *J. Math.* **2019**, *51*, 131–153.
13. Shah, Z.; Bonyah, E.; Islam, S.; Gul, T. Impact of thermal radiation on electrical MHD rotating flow of Carbon nanotubes over a stretching sheet. *AIP Adv.* **2019**, *9*, 015115. [CrossRef]
14. Khan, A.; Nie, Y.; Shah, Z.; Dawar, A.; Khan, W.; Islam, S. Three-dimensional nanofluid flow with heat and mass transfer analysis over a linear stretching surface with convective boundary conditions. *Appl. Sci.* **2018**, *8*, 2244. [CrossRef]
15. Shah, Z.; Bonyah, E.; Islam, S.; Khan, W.; Ishaq, M. Radiative MHD thin film flow of Williamson fluid over an unsteady permeable stretching sheet. *Heliyon* **2018**, *4*, e00825. [CrossRef]
16. Megahed, A. Effect of slip velocity on Casson thin film flow and heat transfer due to unsteady stretching sheet in presence of variable heat flux and viscous dissipation. *Appl. Math. Mech.* **2015**, *36*, 1273–1284. [CrossRef]
17. Shah, Z.; Islam, S.; Ayaz, H.; Khan, S. Radiative heat and mass transfer analysis of micropolar nanofluid flow of Casson fluid between two rotating parallel plates with effects of Hall current. *J. Heat Transf.* **2019**, *141*, 022401. [CrossRef]
18. Jawad, M.; Shah, Z.; Islam, S.; Bonyah, E.; Khan, A.Z. Darcy-Forchheimer flow of MHD nanofluid thin film flow with Joule dissipation and Navier's partial slip. *J. Phys. Commun.* **2018**, *2*, 115014. [CrossRef]
19. Shah, Z.; Islam, S.; Gul, T.; Bonyah, E.; Khan, M.A. The electrical MHD and hall current impact on micropolar nanofluid flow between rotating parallel plates. *Results Phys.* **2018**, *9*, 1201–1214. [CrossRef]
20. Khan, A.; Shah, Z.; Islam, S.; Khan, S.; Khan, W.; Khan, A.Z. Darcy–Forchheimer flow of micropolar nanofluid between two plates in the rotating frame with non-uniform heat generation/absorption. *Adv. Mech. Eng.* **2018**, *10*, 1–16. [CrossRef]
21. Khan, A.; Shah, Z.; Islam, S.; Dawar, A.; Bonyah, E.; Ullah, H.; Khan, A. Darcy-Forchheimer flow of MHD CNTs nanofluid radiative thermal behaviour and convective non uniform heat source/sink in the rotating frame with microstructure and inertial characteristics. *AIP Adv.* **2018**, *8*, 125024. [CrossRef]
22. Shah, Z.; Dawar, A.; Islam, S.; Khan, I.; Ching, D.L.C. Darcy-Forchheimer flow of radiative carbon nanotubes with microstructure and inertial characteristics in the rotating frame. *Case Stud. Therm. Eng.* **2018**, *12*, 823–832. [CrossRef]

23. Khan, N.S.; Gul, T.; Islam, S.; Khan, A.; Shah, Z. Brownian motion and thermophoresis effects on MHD mixed convective thin film second-grade nanofluid flow with Hall effect and heat transfer past a stretching sheet. *J. Nanofluids* **2017**, *6*, 812–829. [CrossRef]

24. Tahir, F.; Gul, T.; Islam, S.; Shah, Z.; Khan, A.; Khan, W.; Ali, L. Flow of a nano-liquid film of Maxwell fluid with thermal radiation and magneto hydrodynamic properties on an unstable stretching sheet. *J. Nanofluids* **2017**, *6*, 1021–1030. [CrossRef]

25. Chu, R.C.; Simons, R. Recent development of computer cooling technology. In Proceedings of the International Symposium on Transport Phenomena in Thermal Engineering, Seoul, Korea, 9–13 May 1993; pp. 17–25.

26. Dixit, T.; Ghosh, I. Review of micro-and mini-channel heat sinks and heat exchangers for single phase fluids. *Renew. Sustain. Energy Rev.* **2015**, *41*, 1298–1311. [CrossRef]

27. Xie, X.; Liu, Z.; He, Y.; Tao, W. Numerical study of laminar heat transfer and pressure drop characteristics in a water-cooled minichannel heat sink. *Appl. Therm. Eng.* **2009**, *29*, 64–74. [CrossRef]

28. Srivastava, L.; Srivastava, V.; Sinha, S. Peristaltic transport of a physiological fluid. *Biorheology* **1983**, *20*, 153–166. [CrossRef]

29. Abbasi, F.M.; Hayat, T.; Shehzad, S.A.; Alsaadi, F.; Altoaibi, N. Hydromagnetic peristaltic transport of copper–water nanofluid with temperature-dependent effective viscosity. *Particuology* **2016**, *27*, 133–140. [CrossRef]

30. Abolbashari, M.H.; Freidoonimehr, N.; Nazari, F.; Rashidi, M.M. Analytical modeling of entropy generation for Casson nano-fluid flow induced by a stretching surface. *Adv. Powder Technol.* **2015**, *26*, 542–552. [CrossRef]

31. Hayat, T.; Muhammad, T.; Shehzad, S.; Alsaedi, A. Three-dimensional boundary layer flow of Maxwell nanofluid: mathematical model. *Appl. Math. Mech.* **2015**, *36*, 747–762. [CrossRef]

32. Malik, M.; Khan, I.; Hussain, A.; Salahuddin, T. Mixed convection flow of MHD Eyring-Powell nanofluid over a stretching sheet: A numerical study. *AIP Adv.* **2015**, *5*, 117118. [CrossRef]

33. Nadeem, S.; Haq, R.U.; Khan, Z. Numerical study of MHD boundary layer flow of a Maxwell fluid past a stretching sheet in the presence of nanoparticles. *J. Taiwan Inst. Chem. Eng.* **2014**, *45*, 121–126. [CrossRef]

34. Raju, C.; Sandeep, N.; Malvandi, A. Free convective heat and mass transfer of MHD non-Newtonian nanofluids over a cone in the presence of non-uniform heat source/sink. *J. Mol. Liq.* **2016**, *221*, 108–115. [CrossRef]

35. Rokni, H.B.; Alsaad, D.M.; Valipour, P. Electrohydrodynamic nanofluid flow and heat transfer between two plates. *J. Mol. Liq.* **2016**, *216*, 583–589. [CrossRef]

36. Nadeem, S.; Haq, R.U.; Khan, Z. Numerical solution of non-Newtonian nanofluid flow over a stretching sheet. *Appl. Nanosci.* **2014**, *4*, 625–631. [CrossRef]

37. Shehzad, S.; Hayat, T.; Alsaedi, A. MHD flow of Jeffrey nanofluid with convective boundary conditions. *J. Braz. Soc. Mech. Sci. Eng.* **2015**, *37*, 873–883. [CrossRef]

38. Sheikholeslami, M.; Hatami, M.; Ganji, D. Nanofluid flow and heat transfer in a rotating system in the presence of a magnetic field. *J. Mol. Liq.* **2014**, *190*, 112–120. [CrossRef]

39. Sheikholeslami, M.; Shah, Z.; Shafee, A.; Khan, I.; Tlili, I. Uniform magnetic force impact on water based nanofluid thermal behavior in a porous enclosure with ellipse shaped obstacle. *Sci. Rep.* **2019**, *9*, 1196. [CrossRef]

40. Mahmoodi, M.; Kandelousi, S. Kerosene- alumina nanofluid flow and heat transfer for cooling application. *J. Cent. South Univ.* **2016**, *23*, 983–990. [CrossRef]

41. Shah, Z.; Gul, T.; Islam, S.; Khan, M.A.; Bonyah, E.; Hussain, F.; Mukhtar, S.; Ullah, M. Three dimensional third grade nanofluid flow in a rotating system between parallel plates with Brownian motion and thermophoresis effects. *Results Phys.* **2018**, *10*, 36–45. [CrossRef]

42. Shah, Z.; Gul, T.; Khan, A.; Ali, I.; Islam, S.; Husain, F. Effects of hall current on steady three dimensional non-newtonian nanofluid in a rotating frame with brownian motion and thermophoresis effects. *J. Eng. Technol.* **2017**, *6*, e296.

43. Shah, Z.; Dawar, A.; Kumam, P.; Khan, W.; Islam, S. Impact of Nonlinear Thermal Radiation on MHD Nanofluid Thin Film Flow over a Horizontally Rotating Disk. *Appl. Sci.* **2019**, *9*, 1533. [CrossRef]

44. Saeed, A.; Islam, S.; Dawar, A.; Shah, Z.; Kumam, P.; Khan, W. Influence of Cattaneo–Christov Heat Flux on MHD Jeffrey, Maxwell, and Oldroyd-B Nanofluids with Homogeneous-Heterogeneous Reaction. *Symmetry* **2019**, *11*, 439. [CrossRef]

45. Kumam, P.; Shah, Z.; Dawar, A.; Rasheed, H.U.; Islam, S. Entropy Generation in MHD Radiative Flow of CNTs Casson Nanofluid in Rotating Channels with Heat Source/Sink. *Math. Probl. Eng.* **2019**, *2019*. [CrossRef]

46. Khan, A.S.; Nie, Y.; Shah, Z. Impact of Thermal Radiation and Heat Source/Sink on MHD Time-Dependent Thin-Film Flow of Oldroyed-B, Maxwell, and Jeffry Fluids over a Stretching Surface. *Processes* **2019**, *7*, 191. [CrossRef]

47. Nasir, S.; Shah, Z.; Islam, S.; Khan, W.; Khan, S.N. Radiative flow of magneto hydrodynamics single-walled carbon nanotube over a convectively heated stretchable rotating disk with velocity slip effect. *Adv. Mech. Eng.* **2019**, *11*. [CrossRef]

48. Nasir, S.; Shah, Z.; Islam, S.; Khan, W.; Bonyah, E.; Ayaz, M.; Khan, A. Three dimensional Darcy-Forchheimer radiated flow of single and multiwall carbon nanotubes over a rotating stretchable disk with convective heat generation and absorption. *AIP Adv.* **2019**, *9*, 035031. [CrossRef]

49. Saeed, A.; Shah, Z.; Islam, S.; Jawad, M.; Ullah, A.; Gul, T.; Kumam, P. Three-Dimensional Casson Nanofluid Thin Film Flow over an Inclined Rotating Disk with the Impact of Heat Generation/Consumption and Thermal Radiation. *Coatings* **2019**, *9*, 248. [CrossRef]

50. Sheikholeslami, M.; Bhatti, M. Forced convection of nanofluid in presence of constant magnetic field considering shape effects of nanoparticles. *Int. J. Heat Mass Transf.* **2017**, *111*, 1039–1049. [CrossRef]

51. Monfared, M.; Shahsavar, A.; Bahrebar, M.R. Second law analysis of turbulent convection flow of boehmite alumina nanofluid inside a double-pipe heat exchanger considering various shapes for nanoparticle. *J. Therm. Anal. Calorim.* **2019**, *135*, 1521–1532. [CrossRef]

52. Sheikholeslami, M. CuO-water nanofluid flow due to magnetic field inside a porous media considering Brownian motion. *J. Mol. Liq.* **2018**, *249*, 921–929. [CrossRef]

53. Sheikholeslami, M.; Shamlooei, M.; Moradi, R. Fe_3O_4-Ethylene glycol nanofluid forced convection inside a porous enclosure in existence of Coulomb force. *J. Mol. Liq.* **2018**, *249*, 429–437. [CrossRef]

54. Munkhbayar, B.; Tanshen, M.R.; Jeoun, J.; Chung, H.; Jeong, H. Surfactant-free dispersion of silver nanoparticles into MWCNT-aqueous nanofluids prepared by one-step technique and their thermal characteristics. *Ceram. Int.* **2013**, *39*, 6415–6425. [CrossRef]

55. Esfe, M.H.; Wongwises, S.; Naderi, A.; Asadi, A.; Safaei, M.R.; Rostamian, H.; Dahari, M.; Karimipour, A. Thermal conductivity of Cu/TiO_2–water/EG hybrid nanofluid: Experimental data and modeling using artificial neural network and correlation. *Int. Commun. Heat Mass Transf.* **2015**, *66*, 100–104. [CrossRef]

56. Akhgar, A.; Toghraie, D. An experimental study on the stability and thermal conductivity of water-ethylene glycol/TiO_2-MWCNTs hybrid nanofluid: Developing a new correlation. *Powder Technol.* **2018**, *338*, 806–818. [CrossRef]

57. Keyvani, M.; Afrand, M.; Toghraie, D.; Reiszadeh, M. An experimental study on the thermal conductivity of cerium oxide/ethylene glycol nanofluid: Developing a new correlation. *J. Mol. Liq.* **2018**, *266*, 211–217. [CrossRef]

58. Ranjbarzadeh, R.; Meghdadi, A.H.; Hojaji, M. The analysis of experimental process of production, stabilizing and measurement of the thermal conductivity coefficient of water/graphene oxide as a cooling nanofluid in machining. *J. Mod. Process. Manuf. Prod.* **2016**, *5*, 43–53.

59. Carreau, P.; Kee, D.D.; Daroux, M. An analysis of the viscous behaviour of polymeric solutions. *Can. J. Chem. Eng.* **1979**, *57*, 135–140. [CrossRef]

60. Kefayati, G.R.; Tang, H. MHD thermosolutal natural convection and entropy generation of Carreau fluid in a heated enclosure with two inner circular cold cylinders, using LBM. *Int. J. Heat Mass Transf.* **2018**, *126*, 508–530. [CrossRef]

61. Olajuwon, B.I. Convection heat and mass transfer in a hydromagnetic Carreau fluid past a vertical porous plate in presence of thermal radiation and thermal diffusion. *Therm. Sci.* **2011**, *15*, S241–S252. [CrossRef]

62. Hayat, T.; Asad, S.; Mustafa, M.; Alsaedi, A. Boundary layer flow of Carreau fluid over a convectively heated stretching sheet. *Appl. Math. Comput.* **2014**, *246*, 12–22. [CrossRef]

63. Alsarraf, J.; Moradikazerouni, A.; Shahsavar, A.; Afrand, M.; Salehipour, H.; Tran, M.D. Hydrothermal analysis of turbulent boehmite alumina nanofluid flow with different nanoparticle shapes in a minichannel heat exchanger using two-phase mixture model. *Phys. A Stat. Mech. Its Appl.* **2019**, *520*, 275–288. [CrossRef]

64. Azari, A.; Kalbasi, M.; Rahimi, M. CFD and experimental investigation on the heat transfer characteristics of alumina nanofluids under the laminar flow regime. *Braz. J. Chem. Eng.* **2014**, *31*, 469–481. [CrossRef]

65. Munir, A.; Shahzad, A.; Khan, M. Convective flow of Sisko fluid over a bidirectional stretching surface. *PLoS ONE* **2015**, *10*, e0130342. [CrossRef] [PubMed]

66. Olanrewaju, P.; Adigun, A.; Fenwa, O.; Oke, A.; Funmi, A. Unsteady free convective flow of sisko fluid with radiative heat transfer past a flat plate moving through a binary mixture. *Therm. Energy Power Eng.* **2013**, *2*, 109–117.

67. Khan, M.; Munawar, S.; Abbasbandy, S. Steady flow and heat transfer of a Sisko fluid in annular pipe. *Int. J. Heat Mass Transf.* **2010**, *53*, 1290–1297. [CrossRef]

68. Khan, M.; Shahzad, A. On boundary layer flow of a Sisko fluid over a stretching sheet. *Quaest. Math.* **2013**, *36*, 137–151. [CrossRef]

69. Patel, M.; Patel, J.; Timol, M. Laminar boundary layer flow of Sisko fluid. *Appl. Appl. Math.* **2015**, *10*, 909–918.

70. Darji, R.; Timol, M. Similarity analysis for unsteady natural convective boundary layer flow of Sisko fluid. *Int. J. Adv. Appl. Math. Mech.* **2014**, *1*, 22–36.

71. Siddiqui, A.; Hameed, A.; Haroon, T.; Walait, A. Analytic solution for the drainage of Sisko fluid film down a vertical belt. *Appl. Appl. Math.* **2013**, *8*, 465–480.

72. Khan, M.; Abbas, Q.; Duru, K. Magnetohydrodynamic flow of a Sisko fluid in annular pipe: A numerical study. *Int. J. Numer. Methods Fluids* **2010**, *62*, 1169–1180. [CrossRef]

73. Sarı, G.; Pakdemirli, M.; Hayat, T.; Aksoy, Y. Boundary layer equations and Lie group analysis of a Sisko fluid. *J. Appl. Math.* **2012**, *2012*. [CrossRef]

74. Khan, H.; Haneef, M.; Shah, Z.; Islam, S.; Khan, W.; Muhammad, S. The combined magneto hydrodynamic and electric field effect on an unsteady Maxwell nanofluid flow over a stretching surface under the influence of variable heat and thermal radiation. *Appl. Sci.* **2018**, *8*, 160. [CrossRef]

75. Moallemi, N.; Shafieenejad, I.; Novinzadeh, A. Exact solutions for flow of a Sisko fluid in pipe. *Bull. Iran. Math. Soc.* **2011**, *37*, 49–60.

76. Dawar, A.; Shah, Z.; Islam, S.; Idress, M.; Khan, W. Magnetohydrodynamic CNTs Casson Nanofl uid and Radiative heat transfer in a Rotating Channels. *J. Phys. Res. Appl.* **2018**, *1*, 017–032.

77. Shah, Z.; Dawar, A.; Khan, I.; Islam, S.; Ching, D.L.C.; Khan, A.Z. Cattaneo-Christov model for electrical magnetite micropoler Casson ferrofluid over a stretching/shrinking sheet using effective thermal conductivity model. *Case Stud. Therm. Eng.* **2019**, *13*, 100352. [CrossRef]

78. Khan, N.S.; Zuhra, S.; Shah, Z.; Bonyah, E.; Khan, W.; Islam, S. Slip flow of Eyring-Powell nanoliquid film containing graphene nanoparticles. *AIP Adv.* **2018**, *8*, 115302. [CrossRef]

79. Khan, W.; Pop, I. Boundary-layer flow of a nanofluid past a stretching sheet. *Int. J. Heat Mass Transf.* **2010**, *53*, 2477–2483. [CrossRef]

80. Osiac, M. The electrical and structural properties of nitrogen Ge1Sb2Te4 thin film. *Coatings* **2018**, *8*, 117. [CrossRef]

81. Bahgat Radwan, A.; Abdullah, A.; Mohamed, A.; Al-Maadeed, M. New electrospun polystyrene/Al$_2$O$_3$ nanocomposite superhydrophobic coatings; synthesis, characterization, and application. *Coatings* **2018**, *8*, 65. [CrossRef]

82. Chen, X.; Dai, W.; Wu, T.; Luo, W.; Yang, J.; Jiang, W.; Wang, L. Thin film thermoelectric materials: Classification, characterization, and potential for wearable applications. *Coatings* **2018**, *8*, 244. [CrossRef]

83. Yamamuro, H.; Hatsuta, N.; Wachi, M.; Takei, Y.; Takashiri, M. Combination of electrodeposition and transfer processes for flexible thin-film thermoelectric generators. *Coatings* **2018**, *8*, 22. [CrossRef]

84. Khan, Z.; Shah, R.; Islam, S.; Jan, H.; Jan, B.; Rasheed, H.U.; Khan, A. MHD Flow and Heat Transfer Analysis in the Wire Coating Process Using Elastic-Viscous. *Coatings* **2017**, *7*, 15. [CrossRef]

85. Ullah, A.; Alzahrani, E.; Shah, Z.; Ayaz, M.; Islam, S. Nanofluids thin film flow of Reiner-Philippoff fluid over an unstable stretching surface with Brownian motion and thermophoresis effects. *Coatings* **2019**, *9*, 21. [CrossRef]

86. Bejan, A. Second-law analysis in heat transfer and thermal design. In *Advances in Heat Transfer*; Elsevier: Amsterdam, The Netherlands, 1982; Volume 15, pp. 1–58.

87. Bejan, A. Entropy generation minimization: The new thermodynamics of finite-size devices and finite-time processes. *J. Appl. Phys.* **1996**, *79*, 1191–1218. [CrossRef]

88. Weigand, B.; Birkefeld, A. Similarity solutions of the entropy transport equation. *Int. J. Therm. Sci.* **2009**, *48*, 1863–1869. [CrossRef]

89. Makinde, O.D. Second law analysis for variable viscosity hydromagnetic boundary layer flow with thermal radiation and Newtonian heating. *Entropy* **2011**, *13*, 1446–1464. [CrossRef]
90. Hayat, T.; Khan, M.I.; Khan, T.A.; Khan, M.I.; Ahmad, S.; Alsaedi, A. Entropy generation in Darcy-Forchheimer bidirectional flow of water-based carbon nanotubes with convective boundary conditions. *J. Mol. Liq.* **2018**, *265*, 629–638. [CrossRef]
91. Makinde, O. Thermodynamic second law analysis for a gravity-driven variable viscosity liquid film along an inclined heated plate with convective cooling. *J. Mech. Sci. Technol.* **2010**, *24*, 899–908. [CrossRef]
92. Esmaeilpour, M.; Abdollahzadeh, M. Free convection and entropy generation of nanofluid inside an enclosure with different patterns of vertical wavy walls. *Int. J. Therm. Sci.* **2012**, *52*, 127–136. [CrossRef]
93. Dawar, A.; Shah, Z.; Khan, W.; Idrees, M.; Islam, S. Unsteady squeezing flow of magnetohydrodynamic carbon nanotube nanofluid in rotating channels with entropy generation and viscous dissipation. *Adv. Mech. Eng.* **2019**, *11*. [CrossRef]
94. Feroz, N.; Shah, Z.; Islam, S.; Alzahrani, E.O.; Khan, W. Entropy Generation of Carbon Nanotubes Flow in a Rotating Channel with Hall and Ion-Slip Effect Using Effective Thermal Conductivity Model. *Entropy* **2019**, *21*, 52. [CrossRef]
95. Alharbi, S.; Dawar, A.; Shah, Z.; Khan, W.; Idrees, M.; Islam, S.; Khan, I. Entropy Generation in MHD Eyring–Powell Fluid Flow over an Unsteady Oscillatory Porous Stretching Surface under the Impact of Thermal Radiation and Heat Source/Sink. *Appl. Sci.* **2018**, *8*, 2588. [CrossRef]
96. Liao, S.J. The Proposed Homotopy Analysis Technique for the Solution of Nonlinear Problems. Ph.D. Thesis, Shanghai Jiao Tong University, Shanghai, China, 1992.
97. Liao, S.J. An explicit, totally analytic approximate solution for Blasius' viscous flow problems. *Int. J. Non-Linear Mech.* **1999**, *34*, 759–778. [CrossRef]
98. Liao, S.J. On the analytic solution of magnetohydrodynamic flows of non-Newtonian fluids over a stretching sheet. *J. Fluid Mech.* **2003**, *488*, 189–212. [CrossRef]
99. Abbasbandy, S.; Jalili, M. Determination of optimal convergence-control parameter value in homotopy analysis method. *Numer. Algorithms* **2013**, *64*, 593–605. [CrossRef]

MDPI

St. Alban-Anlage 66

4052 Basel

Switzerland

Tel. +41 61 683 77 34

Fax +41 61 302 89 18

www.mdpi.com

Processes Editorial Office

E-mail: processes@mdpi.com

www.mdpi.com/journal/processes